T0251637

WEAR OF ROCK CUTTING TOOLS
LABORATORY EXPERIMENTS ON THE ABRASIVITY OF ROCK

WEAR OF ROCK CUTTING TOOLS

Laboratory experiments on the abrasivity of rock

H.J.R. DEKETH

Technical University Delft, Section Engineering Geology, Delft

CRC Press

Taylor & Francis Group

Boca Raton London New York

CRC Press is an imprint of the
Taylor & Francis Group, an **informa** business

Published by

A.A. Balkema, P.O. Box 1675, 3000 BR Rotterdam, Netherlands (Fax: +31.10.4135947)
A.A. Balkema Publishers, Old Post Road, Brookfield, VT 05036, USA (Fax: 802.276.3837)

ISBN 90 5410 620 4

Contents

Preface

The need to get a better understanding of the basic processes of wear in rock cutting has resulted into the study "Wear of Rock Cutting Tools". Two inter-related books were produced in close cooperation of the authors. The books highlight different aspects of the study and may be read standing alone as well as a couple. Both studies are mainly directed towards the effects of different rock types (and geological setting) on the processes of tool wear during rock cutting.

"Laboratory Experiments on the Abrasivity of Rock" by Jan Reinout Deketh provides a better understanding of basic wear processes in laboratory rock cutting wear experiments, by which the site investigation for rock cutting projects like dredging, trenching and tunnelling may be tuned.

"Site Investigation for Rock Dredging" by Peter Verhoef provides a general view over the wide range of aspects which are involved in engineering geological site investigation for rock dredging.

At the time I started this work a lot of preliminary investigations were already carried out by and under the supervision of Peter Verhoef. Moreover also during my work into this subject I was supported by him and other people from the Section Engineering Geology at the Faculty of Mining and Petroleum Engineering of Delft University of Technology. I thank Peter Verhoef for all this and for giving me the opportunity to write this book.

My dearest friend, Marina Giezen, supported me during the years and helped me completing this thesis. She contributed to this work directly by performing a study into the wear of bits of rock cutting trenchers. Her study of wear and production of rock cutting trenchers in different types of rock was essential by demonstrating the relevance of the laboratory experiments to a specific example of rock cutting in practice. Her work was made possible by Vermeer Manufacturing Company (Pella, Iowa, USA).

This research has been sponsored by the Dutch Technology Foundation (STW Stichting Technische Wetenschappen).

Jan Reinout Deketh, December 1994

VIII

CHAPTER 1

Introduction

In many civil engineering and mining projects the earth surface has to be remodelled, prepared or removed for the benefits and needs of a specific project. Such projects involve, for example, open pit or underground mines, tunnels, dredging works, trenches etc. Nowadays manpower is aided or replaced by blasting techniques and excavation machines like dredgers, trenchers, roadheaders, tunnel boring machines etc., which all have their specific advantages and applications. A common problem of most excavation machinery, based on mechanical action, is the unknown interaction of cutting tools with varying types of rock and geological settings. The development of more powerful machines and improved technology enables the cutting of increasingly stronger rock. Unexpected high rates of wear of the cutting tools and low production rates resulting in high financial losses and law-suits between contractors, clients and consultants may be the consequence. The financial losses can be so substantial that the existence of major dredging firms could be endangered. In the work by P.N.W.Verhoef (1995) some cases are discussed of dredging projects which had major problems with wear of the cutting chisels which had not been anticipated.

A large number of investigations has been carried out to determine which factors are of influence to tool wear due to rock cutting. It seems, that wear problems are system dependent problems. Therefore studies about wear are only comparable when the wear system and conditions are more or less equal. Many studies look at a part of a wear problem and therefore only describe the factors which affect that part. Most factors that influence wear due to rock cutting can be placed into the following groups.

1. rock material properties (texture, strength, composition, hardness etc.)
2. rock mass properties (structure, inhomogeneities etc.)
3. type of machinery (type of tools, machine cutting principle etc.)
4. choice of machine settings (thrust, r.p.m. of the rotary cutting device etc.)
5. environment (submerged or dry, weather conditions, operator skills etc.)

For example quartz grains are abrasive to cutting tools made of steel but are hardly abrasive to cutting tools made of tungsten carbide. This is the case at room temperature. If temperatures at the wear-flat of a cutting tool become as high as $\pm$ 800 °C, the hardness of tungsten carbide decreases and quartz (and possibly even softer minerals) becomes as abrasive to the tungsten carbide as to steel. High temperatures can develop if rock cutting machines are not powerful enough to penetrate the rock and thus rub or scrape the rock. Therefore, more powerful rock cutting machines are able to work in a more advantageous wear mode (e.g. lower temperatures) in stronger rocks, which results in a lower rate and a different type of wear of the machine cutting tools. These examples show that the rate and type of wear are affected by many different factors, which all should be considered when analyzing or optimizing the rock cutting wear processes.

Wear, due to rock cutting, has been approached from different disciplines. Many of them have studied the type of cutting tool or method of cutting while using only one or a few rock types for testing purposes. However, in practice the variety of rock types in different rock cutting projects may well exceed the variety of rock cutting machinery, machinery settings or environmental conditions.

An approach to study wear processes of cutting tools working in different rock types, is by carrying out laboratory experiments. Laboratory tests can and have been carried out at settings such, that the effect of different variables could be determined rather easily. As laboratory experiments become more simple, they facilitate changes in the variables and therefore can be applied to a wider variety of rock cutting projects. However, laboratory tests may only describe part of a full scale in-situ rock cutting process and therefore can only indicate qualitatively which factors might be of influence to the full scale wear process. On the contrary, complex experiments, like in-situ changes of machine settings in a rock excavation project, describe the wear process very well, but the results can only be used for that specific project, since no two projects are the same.

Tests to obtain parameters, which describe the cuttability and abrasivity of rock, have been developed (Roxborough and Philips 1974, Schimazek and Knatz 1970, 1976, Gehring 1987). Because of many simplifications in these tests with respect to the in-situ wear system, the applicability of the test results as a quantitative measure for the expected amount of wear in-situ, is questionable. However, wear experiments may be helpful to understand how wear can occur under specific conditions. With this knowledge the cutting device may be tuned better, the cutting process may be optimized and the assessment of expected rates of wear may be more reliable.

Before a wear test is designed, it is necessary to determine what wear mechanisms dominate at given conditions. Simplifications to the cutting process are only justified in the experiments, if they do not influence the wear mechanism as such, i.e. the prevailing wear mechanism should not differ from the mechanism occurring in-situ. A thorough study and understanding of the possible cutting and wear mechanisms is therefore needed.

In this work, first a study is carried out to establish which cutting and wear mechanisms prevail during rock cutting (chapter 2 and 3). Based on this a hypothesis (the wear mode hypothesis) is made (chapter 4), which is verified in laboratory experiments. The laboratory test, the "scraping test", used for the experiments (chapter 7 to 11), is described in chapter 5. In chapter 10 tests are performed on artificial rock, which enables to vary the rock properties independently. In chapter 11 test results on natural rock types, sandstones and limestones, are discussed and compared to the results of the experiments on artificial rock. Finally, the experimental results are applied to two rock cutting projects with rock cutting trenchers, as an example of the relevance of the laboratory results to practice of rock cutting (chapter 12).

The *general objective* of this study is to enlarge the understanding of basic wear mechanisms operating during rock cutting, by performing experiments with a small scale laboratory rock cutting test, the "scraping" test. A better understanding of basic wear mechanisms may benefit the optimization of rock cutting tools, improve the prediction of wear in rock cutting projects and supply for the necessary framework of knowledge to which other investigations into this topic can be tuned.

This study is focused to wear processes occurring in the range of feed of a chisel into the rock, where a transition from a scraping to a cutting process takes place. The independent variation of rock properties in the experiments is emphasized. The steel type of the test chisels and the cutting velocity has been varied as well. The experiments have been performed in order to:

- show that an increase of the feed of the chisel into the rock can lead to a change of mode (type and rate) of wear.
- show that this behaviour is caused by the occurrence of different wear modes as described in the wear mode hypothesis.
- investigate the influences of some rock material properties on the different wear modes.
- investigate the effects of the cutting velocity on the rate and type of wear.
- investigate the influences of some rock properties on the type and rate of wear at different cutting velocities.

Next to investigations of rate and type of wear as described in the previous objectives, the cuttability of different types of rock is considered.

Questions, which remain after this study, are:

- which criteria should be used to assess the parameters, which characterize the wear capacity of rock, in thin section analysis?

- what information is needed to apply the results from this study to rock cutting projects in practice quantitatively?

CHAPTER 2

Rock cutting theories

Many rock cutting theories have been developed in the past decennia. In most cases these theories apply to tunnelling (and not to dredging or trenching). A data-bank search specifically in ground excavation carried out at the U.S. Army Waterways Experiment Station, Vicksburg, Mississippi, produced over 25,000 references, over a thousand pertaining to rock excavation. A similar search on rock cutting dredging produced less than a dozen references (Hignett 1984). Despite the fact that there are many similarities between rock excavation by dredging and tunnelling, there are also large differences, which make direct application of rock tunnelling research to rock cutting dredging or trenching questionable. Besides, each research into rock cutting is very specific and resulting rock cutting models are often not applicable to other situations. However, a brief impression of some results of various authors, which investigated rock cutting processes, is given below.

During many rock cutting experiments, the formation of chips of rock material in front of chisels, picks, bits or wedges has been observed. Dalziel and Davis (1964) and Bisschop (1991) measured cutting forces during rock cutting experiments. Diagrams of the cutting forces versus the time show a zigzag pattern. When a large chip is formed, the cutting forces reach a maximum. After the formation of the large chip, the cutting forces quickly drop to a minimum value, then steadily increase until a new large chip is formed and again a maximum of the cutting forces is reached. Koert (1981) observed also smaller rock chips breaking out during the build-up phase of the cutting forces in his cutting experiments.

The formation mechanism and the shape of rock chips might be determined by factors like rock material (and rock mass) properties, shape of the cutting tool and depth of cut (Clark 1987). Also the forces necessary to cut the rock are determined by these factors, together with the state of stress of a rock and the presence of free surfaces to which fractures induced by the cutting tool may propagate (Larson, Morell and Mades 1987).

Merchant's theory (1944) postulates that the chips are formed by shear failure. His theory is based on ductile behaviour of the rock.

Evans (1962) compared Merchant's theory with laboratory tests on four different types of coal. He did not find any signs of plastic deformation on the failure planes of the chips and concluded that the chips were not formed by shear failure and the rock behaved as a brittle material. He proposed his own theory, which considers that

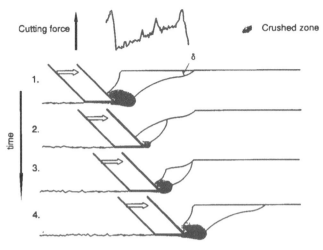

Figure 2 The chip size determines the cutting forces and the size of the crushed zone. The momentary cutting depth of the chisel varies with the variation of size and shape of the chips and the crushed zone.

the formation of chips is essentially caused by tensile failure. Roxborough (1973) validated the findings of Evans for rock types other than coal, which gives Evans rock cutting model a wider application. B.N. Whittaker and A.B. Swilzski (1973) remarked, that the chips in the experiments of Evans had a conchoidal shape.

According to Y. Nishimatsu (1971), failures are induced as a result of compressive and tensile forces. He stated that chips are formed in brittle failure.

In addition to the formation of chips, part of the rock material might be crushed in front of the cutting tool tip (Nishimatsu 1971, Clark 1987). This zone of crushed material exerts a stress upon the rock material and causes cracking of the rock material, which leads to the formation of chips.

Theories of Delft Hydraulics are also based on the existence of a crushed zone (Delft Hydraulics pers. comm.). Camera monitored experiments confirmed the existence of the crushed zone. The formation of rock chips is illustrated in Figure 2. The shape and size of the chips depend on the enclosure angle and the size of the crushed zone. The enclosure angle δ is the angle between the free rock surface and the fracture plane, which originates at the rake face of the chisel near the border of the crushed zone and ends where it meets the free rock surface. The cutting forces may vary according to the formation of minor or major chips.

Cools (1993) carried out rock cutting experiments on limestone and sandstone, both of about the same strength (25-30 MPa). Surprisingly, the chisel (a full scale ESCO pick-point chisel) was extremely worn after cutting the limestone, whereas wear was negligible after cutting the sandstone. The temperatures on the wear-flat of the chisel, the cutting forces and wear phenomena on the chisel wear-flat were different for the two rock types. The chisel showed parallel tracks, polished edges and crushed rock fragments bedded into the wear-flat of the chisel after testing the sandstone and plastic deformation, especially near the face of the chisel, after testing the limestone. The cutting forces and the temperatures measured were much higher while testing the limestone.

A possible explanation for the difference in cutting action and wear could be that the limestone behaved in a ductile fashion whereas the sandstone behaved in a brittle fashion during cutting. If the stress at the wear-flat of the chisel is higher than the brittle-ductile transition-stress, the rock behaves in a ductile way. Since the brittle-ductile transition-stress of the limestone was exceeded by the stress applied by the chisel to the rock, the rock behaved in a ductile way during the cutting experiment, whereas the brittle-ductile transition-stress of the sandstone was about equal to the stress applied by the chisel on the rock and behaved, therefore, probably in a brittle manner.

In most rock cutting machinery, rock cutting tools such as picks, chisels and roller cutters act within in an array of many identical tools. The geometry of such an array and the spacing between the tools affect the cutting process. Roxborough (1973) found that interaction of the tools increases the cutting efficiency. The interaction between the cutting tools is optimal at a specific spacing of the cutting tools, defined by the depth of cut and the side break-out or side-splay angle of the rock.

The cutting theories discussed so far are concerned with the cutting of rock materials and do not take the influence of rock mass characteristics, like discontinuities[1], into account. Discontinuities act as free rock faces which affect the crack propagation in front of a cutting tool. The influence of the discontinuities on the cutting process becomes significant if the spacing between the discontinuities is in the same order of magnitude as the cutting depth of the cutting tool. Various authors have incorporated different rock mass property parameters into a rating system for digging, ripping or cutting purposes. Franklin et al. (1971) for example used the joint density and the unconfined compressive strength of the rock to determine the excavatability[2] of a rock mass. Weaver (1975, cited by Bell 1992) developed a rippability chart based upon Bieniawski's rock mass classification system, and added a rating for the seismic velocity. Kirsten (1982, cited by Bell 1992) produced an excavatability index based upon Barton's rock mass classification system. However, it is unlikely that these rating systems are applicable to the wide spectre of existing rock cutting machinery or to the total range of working conditions of such machinery. MacGregor (1993) compared different rating systems by monitoring rippers working under different geological conditions. The relation of the rating systems with performance of the rippers was often poor.

Despite the apparent importance of rock mass characteristics on many rock cutting, digging or ripping operations, this will not be discussed in this work since elaboration about these aspects would cloud rather than clarify its contents. P.N.W. Verhoef (1995) treats this subject extensively in "Wear of Rock Cutting Tools, Site Investigation for Rock Dredging".

---

[1] According to British standard (BS 5930 : 1981) discontinuities are fractures in the rock mass and include joints, fissures, faults, shear planes, cleavages and bedding.

[2] Cuttability, excavatability, rippability refer to the facility of rock to be excavated by cutting tools. The excavating process is respectively determined by intact rock material properties, by the rock mass properties (material properties and discontinuities) and only by discontinuities.

CHAPTER 3

Possible wear mechanisms at the wear-flat of rock cutting tools

3.1 INTRODUCTION

The process of rock cutting involves different types of motion contact between rock and tool, causing different kinds of damage to the cutting tool. According to Zum Gahr (1987) the mechanisms causing damage are plastic deformation, corrosion, cracks and wear. Only wear is considered in this study. According to DIN 50320 (in Zum Gahr 1987) wear is defined as "the progressive loss of material from the surface of a solid body due to mechanical action, i.e. the contact and relative motion against a solid, liquid or gaseous counterbody". Wear of rock cutting tools mainly occurs at their wear-flats (Davids and Adrichem 1990). In Figure 3 a pick-point of a cutter suction dredger is shown as an example of such a cutting tool.

The difference in motion contact as well as the properties of tool, rock and the environment may affect the wear process. To describe the different wear processes, wear type classifications are made. In this study, the wear-type classifications described by Zum Gahr (1987) have been used as a basis to determine what types of wear might be relevant to the wear of rock cutting tools.

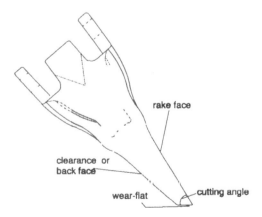

Figure 3 An example of a pick-point (ESCO type 54D 16B) of a cutter suction dredger

7

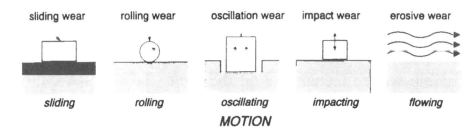

Figure 4 Motion categories to classify wear after Zum Gahr (1987).

3.2 WEAR TYPE CLASSIFICATIONS

To describe wear, wear-type classifications are made. They are based on terms related to features of worn surfaces, types of motion or wear mechanisms. In his book "Microstructure and Wear of Materials" Zum Gahr (1987) extensively describes the subject matter. A motion category classification and a wear mechanism classification, are used in this study to classify wear in mechanical rock cutting operations. The different types of wear distinguished in the wear-type classifications are ideal situations. In practice the type of wear is often a mix of several ideal wear-types, in which some may be more important than others. Many parameters, which are often interdependent, affect the final type of wear. Such factors are for example the nature of the involved materials (brittleness, hardness), the temperature and contact stresses and the type of contact motion (e.g. sliding, impact).

3.2.1 *The motion category classification*

The categories of motion, described by Zum Gahr, are: rolling, sliding, oscillating, impact and flowing (Figure 4). The wear of rock cutting tools (chisels, picks and bits) is caused mainly by impact and sliding motion at the tool rock interface.

Impact directly causes wear by cracking or flaking and indirectly by weakening the microstructure of the tool material.

In this research wear due to sliding is studied. Sliding motion of a rock cutting tool over a rock surface causes sliding wear or grooving wear. *Grooving wear* is caused by abrasion as the main wear mechanism whereas in *sliding wear* elastic and plastic deformation, adhesion and surface fatigue are the main mechanisms.

3.2.2 *The wear mechanism classification*

According to DIN 50320 (in: Zum Gahr 1987) wear, due to a sliding motion, may be the result of four basic wear mechanisms: adhesion, abrasion, tribochemical reaction and surface fatigue (Figure 5).

Plastic deformation as such is generally not called a wear mechanism, but it plays an important role in many wear processes.

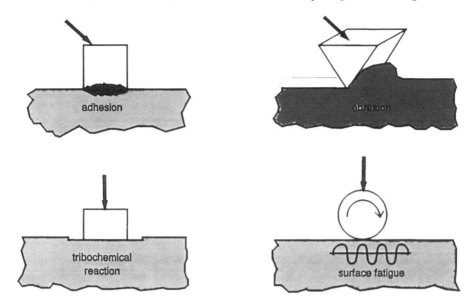

Figure 5 Four basic wear mechanisms (after Zum Gahr (1987))

Wear due to *surface fatigue* can be characterized by crack formation and flaking of material caused by repeated alternated loading of solid surfaces. Repeated sliding contact of asperities on the surfaces of solids in relative motion may result in surface fatigue on a microscopic scale.

Wear by *tribochemical reaction* can be characterized by "rubbing" contact between to solid surfaces that react with a gaseous or liquid environment. The wear process takes place by continuous removal and new formation of reaction layers on the contacting surfaces.

Adhesive wear is defined as wear due to adhesive material transfer. High local pressure between contacting asperities results in plastic deformation, adhesion and consequently the formation of local junctions. Relative sliding between the contacting surfaces causes rupture of these junctions and frequently the transfer of material from one surface to another.

Abrasive wear[3] is defined as the displacement of material caused by the presence of hard protuberances on the rock surface (two-body wear) or by hard particles situated between the two sliding surfaces (three-body wear). In Figure 6, two- and three-body wear systems are illustrated.

Abrasion can be divided into four types of material failure: microploughing, microcutting, microfatigue and microcracking (Figure 7). Microploughing, microcutting and microfatigue are the dominant types of material failure in more ductile materials such as steel. In the ideal case, microploughing due to a single pass of one abrasive particle does not result in any detachment of material from a wearing

---

[3] Abrasive wear is the main mechanism in grooving wear. The term grooving wear can therefore be used to describe the same type of wear as abrasive wear.

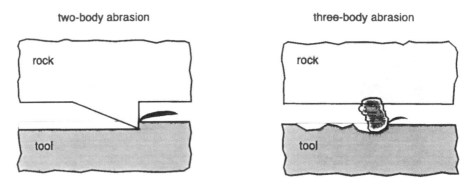

Figure 6 Two- and three-body wear (after Zum Gahr 1987)

surface. A prow is formed ahead of the abrading particle and material is continually displaced sideways to form ridges adjacent to the groove produced. During microploughing, material loss can however occur due to many abrasive particles which are acting simultaneously or successively. Material may be ploughed aside repeatedly by passing particles and may break of by fatigue. Microcracking is related to brittle materials like tungsten carbide.

3.3 WEAR MECHANISMS RELEVANT TO ROCK CUTTING TOOLS

Abrasive (or grooving) wear and adhesive wear (due to high temperatures and contact stresses) are assumed to dominate the wear process during rock cutting of rock types containing minerals harder than the tool material. For example quartz containing rock types cause abrasive wear to steel tools, often accompanied by adhesive wear if the temperatures and stresses are high enough. Surface fatigue and tribochemical reaction are only considered to play a role if the rate of wear is rather low, which allows for the necessary time for these processes to take place.

If the mineral grains are softer than the material of the cutting tool during the cutting process, sliding wear[4] is likely to occur[5]. Sliding wear can occur during the cutting of rock types, bearing only relatively soft minerals, like the cutting of calcarenite by dredge pick-points or the cutting of sandstones by tungsten carbide picks provided that the relative hardness of the tool material remains higher than the hardness of the rock minerals during the cutting process.

---

[4] Sliding wear and grooving wear are both a product of sliding motion contact of two bodies with or without an interfacial element (third body).

[5] Minerals, which are known to be relatively soft at room temperature, may be relatively hard at higher temperatures due to loss of hardness of the tool material at higher temperatures. Therefore, although sliding wear is expected during rock cutting, based upon the relative hardness of minerals and tool material at room temperature, abrasive wear occurs during cutting, due to a change of relative hardness.

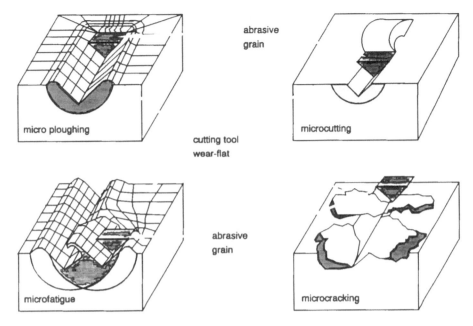

abrasive
grain

cutting tool
wear-flat

abrasive
grain

Figure 7 Four types of material failure (after Zum Gahr 1987)

3.3.1 *Abrasive wear*

Abrasive wear is caused by sliding of a surface (tool wear-flat) over hard asperities of another surface (rock surface) (two-body wear) or by hard loose particles between two opposite sliding surfaces (three-body wear). Two-body wear causes wear one to two orders higher than three-body wear (Mishra and Finnie 1979, Loes 1971, Zum Gahr 1987).

The resistance of the asperities embedded in the rock surface to crushing due to a rock cutting tool sliding over the asperities, determines whether two or three-body wear will dominate the wear process. An increase of the resistance to crushing of the asperities will favour two-body wear. Contact stresses will be high when the asperities remain intact. High contact stresses force the asperities deep into the steel tool, thus causing a lot of abrasive wear. High temperatures may arise at the contacts between rock and tool, which may affect the relative hardness of rock mineral and tool material, thus altering the range of minerals in the rock supposed to be abrasive. The different modes of response of the rock surface to loading by a rock cutting chisel is illustrated in Figure 8. In wear mode I the asperities remain intact and in wear mode III the asperities fail by the action of the chisel. Wear mode II represents the transition from wear mode I to wear mode III. A description of the three wear modes is given in Figure 8.

A subdivision of three-body wear can be made according to Uetz (1986), based on mineral grain hardness and rock matrix hardness (strength) relative to material hardness of the cutting tool:

1. the mineral grains are harder than the tool material and the weaker rock matrix.

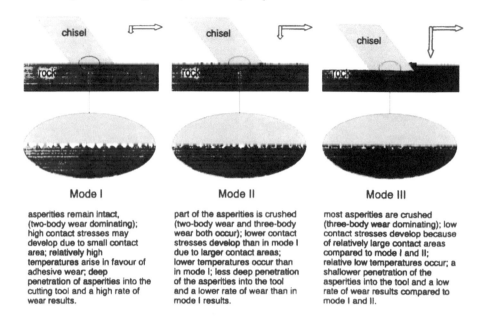

Mode I

asperities remain intact, (two-body wear dominating); high contact stresses may develop due to small contact area; relatively high temperatures arise in favour of adhesive wear; deep penetration of asperities into the cutting tool and a high rate of wear results.

Mode II

part of the asperities is crushed (two-body wear and three-body wear both occur); lower contact stresses develop than in mode I due to larger contact areas; lower temperatures occur than in mode I; less deep penetration of the asperities into the tool and a lower rate of wear than in mode I results.

Mode III

most asperities are crushed (three-body wear dominating); low contact stresses develop because of relatively large contact areas compared to mode I and II; relative low temperatures occur; a shallower penetration of the asperities into the tool and a low rate of wear results compared to mode I and II.

Figure 8 Mode I, II and III. The arrows indicate the rate of displacement (feed) perpendicular and parallel to the rock surface

Hard abrasive grains will be pressed and bedded into the weaker rock matrix. Tops of the mineral grains stick out of the rock surface as abrasive asperities.

2.the mineral grains and rock matrix are relatively weak compared to the hardness of the cutting tool material. Wear by abrasion is not likely to occur.

3.the mineral grains and rock matrix are relatively hard compared to the hardness of the cutting tool material. The hard abrasive mineral grains are pressed and bedded into the cutting tool material and form a protective layer against abrasive action from the rock surface.

The material failure mechanisms of abrasive wear are determined by the shape of the asperities, tool material properties and whether two- or three-body wear prevails.

Microcutting is thought to be the main mechanism behind two-body wear, whereas three-body wear is related to the mechanism of microploughing[6] (if steel cutting tools are considered).

Angular asperities cause tensile stresses in the tool surface layer whereas rounded asperities induce compressive stresses in the tool surface layer. Iverson (1991) studied the abrasion of glacier beds by rock fragments bedded in the glacier and found that compressive forces prevail when the attack angle is small and tensile forces will prevail when the attack angle is large (Figure 9).

---

[6] Mishra and Finnie (1981) compared two-body wear with small angle erosion, where material is mostly removed by a metal-cutting process (microcutting) and three-body wear with large angle erosion where rather material deformation takes place (microploughing).

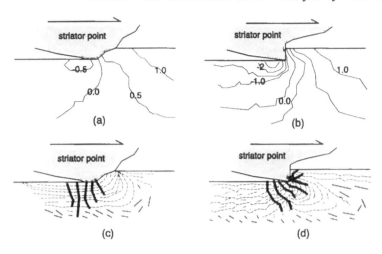

Figure 9 Abrasion of glacier beds by rock fragments (Iverson (1991). Tensile stress magnitudes and orientations in the glacier bed beneath ploughing striator points. Normalized tensile stress contours, (a) and (b). Trajectories of tensile stress and failure paths (indicated by the thick lines), (c) and (d). Attack angles of (a),(c) and (b),(d) are respectively 45° and 90°

There are different failure mechanisms for different shapes of the asperities, probably related to differences in stress regime. Asperities with a small attack angle cause abrasive wear by microploughing and asperities with a large attack angle are related to microcutting. In Figure 10 the attack angle "α" is related to the ratio of microcutting to microploughing. The critical attack angle α_c is determined by the angle at which microcutting and microploughing are contributing equally to the total wear process. The critical attack angle is a function of the test conditions and the wearing (cutting tool) material (Zum Gahr 1987).

The hardness and brittleness of the cutting tool, which are interrelated, influence the prevailing material failure mechanisms and the abrasive wear resistance of the tool as well (Figure 11). The wear resistance changes with the occurrence of different abrasive failure mechanisms. If the hardness of a material increases generally its brittleness increases as well. Cutting tools made of hard, brittle materials subject to abrasive wear, like tungsten carbide picks abraded by a rock containing minerals with a Mohs hardness greater than 7, will wear by microcracking of the tool material surface. A cutting tool made of softer, more ductile material, like steel used for rock dredging pick-points, which cuts the same rock type will wear away by microploughing and microcutting of the tool material surface. In Figure 12 results from pin abrasion tests (Zum Gahr 1987) are displayed. The hardness of the tool material affects not only the abrasive wear resistance but also the material failure mechanism. Ceramics show lower abrasion resistance than pure metals of comparable hardness when hard abrasives are used. However, next to hardness, other factors affect the wear resistance as well.

Other factors, which influence abrasive wear can be divided into two groupes: a group of factors controlling the cutting forces and a group of factors which influence

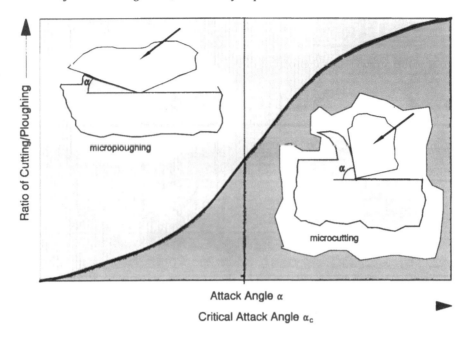

Figure 10 Ratio of microcutting to microploughing relates to the attack angle of the abrasive asperity (after Zum Gahr 1987)

the surface response of the rock material to the rock cutting tool. These factors are discussed in chapter 4.

3.3.2 *Adhesive wear*

Adhesive wear is defined as wear due to adhesive material transfer. High local pressures between contacting asperities result in plastic deformation, adhesion and consequently the formation of local junctions. Relative sliding between the contacting surfaces causes rupture of these junctions and frequently the transfer of material from one surface to another. High temperatures may arise at the wear-flat due to friction between rock and tool. As already discussed in the previous paragraph high temperatures may occur in wear mode I and adhesive wear will take place, additional to abrasive wear. Factors, which may influence adhesive wear, are:
    1. The cutting velocity.
An increase of the cutting velocity causes an increase of the temperatures at the wear-flat of the rock cutting tool. (Schimazek 1970)
    2. The geometry of the cutting tool.
A blunt chisel, for example, will cut the rock with greater difficulty than a sharp one (Kenny and Johnson 1976). Extra energy (forces) is needed for a blunt chisel to cut an equal volume of rock (Roxborough and Philips 1981). More rock material will fail by crushing and temperatures will be higher in comparison with a sharp chisel. Adhesive wear will take place or increase with increasing bluntness.

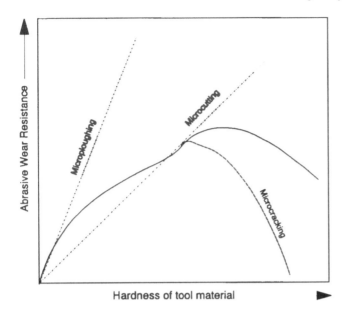

Figure 11 Schematic representation of abrasive wear resistance as a function of hardness (after Zum Gahr 1987)

3.The conductivity and temperature sensitivity of the tool material.

If the conductivity of the tool material is higher, heat can dissipate faster and temperatures at the wear-flat remain lower. For example steel types, which are sensitive to high temperatures, are more susceptible to adhesive wear. (Colijn pers.comm.[7]). Muro (1978) carried out pin-on-disc type experiments and observed a more rapid increase of the rate of wear above a certain normal load on the steel pin. He suggested that this increase was caused by high temperatures, which softened the steel.

4.Mechanical properties of the rock.

An increase of the resistance to crushing and chipping of the rock causes an increase of the cutting forces. An increase of the cutting forces may cause an increase of contact-stresses and thus an increase in friction, temperature and wear. (Whitbread (1960) in Osburn 1969). If the asperities of the rock relief surface fail due high contact-stresses, which exceed the strength of the asperities, other wear mechanisms may occur. This may result in a different wear mode, causing a different type and a different rate of wear.

5.Texture and composition of the rock.

The friction between the rock cutting tool and the rock surface depends on pressures and on the morphology of the rock relief. A rough relief will probably cause more friction and thus higher temperatures than a smooth relief. The roughness of a relief and its strength may depend on the texture and composition of the rock.

---

[7] Faculty of Chemistry and Material Sciences, Delft University of Technology.

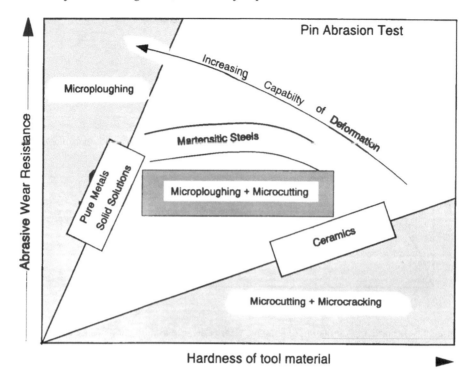

Figure 12 Schematic representation of abrasive wear resistance against a hard abrasive versus bulk hardness of (tool) materials (after Zum Gahr 1987)

3.3.3 *Sliding wear*

Abrasive (or grooving) wear occurs if the abrasive is harder than the wearing body (cutting tool). Abrasion of materials by soft abrasives occurs by rubbing, (sliding wear). Abrasive particles may be called soft, when their hardness is equal to or less than the hardness of the material of the wearing body. The attack of relatively soft abrasives may result in elastic and plastic deformation, adhesion and surface fatigue. The wearing body (tool) failure mechanism is surface cracking and protuberances, due to plastic deformation of the rubbed surface, may be cut or repeatedly pushed aside by following soft abrasive particles. If high temperatures arise at the wear-flat adhesive wear will become the main wear mechanism.

3.3.4 *Total wear due to sliding motion contact*

Abrasive wear and adhesive wear make up the greater part of the total wear due to sliding of a chisel wear-flat over abrasive rock surfaces.

Adhesive wear contributes to the total wear when the temperature and the contact stresses are high enough to weaken the cutting tool material and when the cutting tool

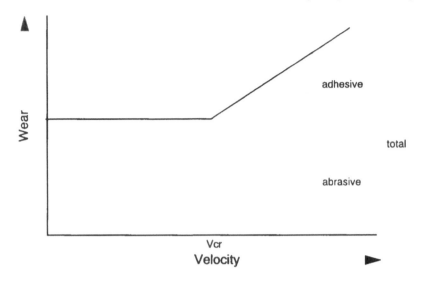

Figure 13 A possible build-up of the total wear due to sliding motion

is worn by soft abrasives (softer than the cutting tool material). The temperature at which the steel first starts to weaken is called the critical temperature (T_{cr}). An increase of the cutting velocity may cause an increase of temperature. The velocity, at which the critical temperature is reached, is called the critical velocity (V_{cr}) (Schimazek 1976, Sellami 1993). The critical velocity is affected by many factors such as cutting tool-geometry, the tool material and rock properties. Figure 13 illustrates how abrasive and adhesive wear may contribute to the total wear.

Obviously different wear mechanisms dominate at different operating conditions. This idea is the basis of the wear mode hypothesis proposed in chapter 4.

CHAPTER 4

Wear mode hypothesis

4.1 INTRODUCTION

In the initial stage of this research project a wear mode hypothesis was developed, which will be outlined in this chapter. This hypothesis has been tested by experiments, which will be treated in chapter 7 to 11. At this stage of the research, the model proves to be a valuable aid in understanding the wear processes[8] operating during rock cutting. The hypothesis is based on consideration of the rock surface structure (relief) over which a cutting tool is sliding during the cutting action. (Wear caused by impact of a cutting tool on a rock surface is not considered). Crushing of the rock surface relief may be caused when cutting forces cause such high stresses on the relief asperities that they will fail (Figure 8, chapter 3). When the rock surface relief changes, the abrasive action of the rock may change. Zum Gahr (1987) and Malkin (1989) distinguish two modes of abrasion; high stress abrasion and low stress abrasion. The stresses at which the asperities of the rock fail are related to rock material properties (among other parameters), as are the cutting forces. With an increase in the displacement rate of the cutting tool into the rock, the stresses will increase until the rock relief fails. At this point a change of cutting mode and wear process or wear mode takes place.

4.2 CUTTING FORCES

Many factors influence wear in rock cutting operations. Most of them belong to one of the following groups:
   1. rock material properties (texture, composition, strength, hardness etc.)
   2. rock mass properties (structure, discontinuities, inhomogeneities etc.)
   3. type of machinery (type of cutting tools, machine cutting principle etc.)
   4. machinery settings (thrust, condition of the tools at replacement etc.)
   5. environment (submerged or dry, weather conditions etc.)
The factors which influence rock cutting forces are divided into three main groups:

---

[8] The wear mode model only considers grooving wear; two- or three-body abrasive wear by hard abrasives of a rock cutting tool in sliding motion contact with the rock.

a group concerning rock material property factors (group 1), a group concerning rock mass properties (group 2) and another group concerning operating condition factors (group 3 to 5).

4.2.1 *Influences of rock material properties on cutting forces*

Resistance of rock material to cutting or cuttability of rock may determine the magnitude of the cutting forces: the shear or cutting force and the normal force. Rock material properties such as strength, stiffness and hardness among others determine the resistance of rock to cutting. These properties are determined by composition and texture of the rock.

4.2.2 *Influences of rock mass properties on cutting forces*

The cutting forces result from the resistance of rock to cutting. However, rock resistance to cutting can be divided into rock material and rock mass resistance. In many cases rock mass resistance to cutting will be lower than rock material resistance due to structural defects. Discontinuities determine rock cutting resistance by their frequency and orientation (Hoogenbrugge 1980, Braybrooke 1988, Fowell 1991).

4.2.3 *The influence of some operating condition factors on cutting forces*

Vibrations in the cutting device may reduce the cutting forces. Koert (in van Rossen 1988) has carried out rock cutting experiments with such vibrations. The cutting forces, measured during these tests, were only 50% of the cutting forces during tests without any vibration.

According to van Rossen (1988) water plays an important role with respect to submerged rock cutting mechanisms. Cutting forces increase due to dilatation of the rock. As rock chips break out, cracks form and dilate. In the space of the dilating cracks the pressure is low which will induce water to flow into the cracks. The relatively low pressure in the cracks is partly increased by the water inflow until an equilibrium is reached between the pressure difference and the resistance of the water inflow. This mechanism leads to an increase in the magnitude of the cutting forces.

The geometry of the cutting tool and the attack angle influence the cutting forces (Roxborough and Rispin 1973, Roxborough and Sen 1986). Blunted chisels require more force to cut the rock (Dalziel and Davies 1964). An increase of the attack angle causes a decrease in the cutting forces. Blunted chisels caused high cutting forces during rock cutting experiments performed by Kenny and Johnson (1976).

Interaction of cutting tools during the cutting process reduces the cutting forces dramatically (Friedman and Ford, 1983). The interaction depends on the spacing and cutting depth of the tools (Roxborough and Sen 1986).

With an increase in cutting depth, the cutting forces increase or remain constant (Kenny and Johnson 1976).

The cutting forces during rock cutting with very shallow cutting depths, as in the scraping test described in chapter 5, do not show a saw-tooth pattern (Figure 2), since no (large) rock chips are formed.

Rock shows a higher strength with an increase of the rate of loading of the rock (Clark 1987). Therefore cutting forces, related to the strength of the rock, will increase with increasing cutting velocity (Gregor 1968 in Nishimatsu 1979). Other authors concluded that cutting forces are independent of tool velocity, since the velocity of crack propagation is much larger (factor 100) than the tool velocity (Nishimatsu 1979). Roxborough (1973) found that the cutting tool velocity had no significant effect on the cutting forces in cutting tests on anhydrite. He also stated that the cutting forces would have increased with increase of cutting tool velocity if abrasive rocks would have been tested. In this case the cutting forces increase due to increase of blunting of the cutting tools by the abrasive rock with increase of cutting velocity.

4.3 ROCK RELIEF STRENGTH

The relief strength is defined as the resistance to crushing of a rock surface relief. The relief strength is determined by rock material strength and relief geometry, which in their turn depend on rock composition and texture. When load is applied to the rock by the cutting tool, the relief will deform, first elastically, than plastically and finally fails. As the relief deforms the contact area between rock and wear-flat enlarges. With an increase of the deformation or contact area, the stiffness of the relief will increase (Hopkins 1990) and the stresses decrease. The relief deforms until the applied load becomes equal to the force, resisting deformation (Mecklenburg and Benzing 1976).

The relief strength may depend on many factors, which can be grouped into rock material strength properties, relief geometry and the strength or hardness of the asperities. These factors are influenced by the composition and texture of the rock.

4.3.1 *The influence of rock material strength properties on relief strength*

1.The strength of rock material
The rock asperities consist of rock material and therefore their strength may be related to the rock material strength. However, the asperities may consist of individual grains of a different strength than the rock material strength and therefore their strength may also be independent of the rock material strength. In both cases the stresses, imposed on the asperities by the cutting tool, are distributed into the rock beneath the asperities. The rock beneath the asperities reacts upon loading according to its material strength (which may be a function of the grain frame and the binding matrix for sedimentary rock types) and thereby rock relief strength is influenced by rock material strength.

2.The stiffness of rock material
Rock is composed of different minerals. These minerals all have a different strength and stiffness, both being anisotropic in most cases. The average stiffness of the different minerals present, determines the stiffness of the rock material. If the amount

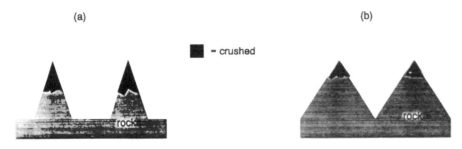

Figure 14 Steep sharp asperities are more sensitive to crushing than broad ones

of stiff minerals is so large that they form a skeleton, the less stiff minerals do not influence the stiffness of the rock material any more. The stiffness of the relief is determined by the stiffness of the rock material and the geometry of the relief.

3. The effect of strain rate on rock stiffness and strength

Rock stiffness as well as rock strength increase with an increase in strain rate. In standard rock material strength tests, the strength determined at very high strain rates (impact) may be a factor two larger than the strength determined at lower strain rates (Reinhardt 1985). Above a certain strain rate the effective strength does not increase further. (Clarke 1987).

4.3.2 *The influence of the geometry of the relief on relief strength*

The shape of the relief asperities, as well as their distribution may affect the strength of the rock surface.

The *shape* of the asperities will probably affect their strength, because the shape also affects the strength of rock cores (Farmer 1983). A large length-diameter ratio of rock cores results in a lower compressive strength (chapter 6). Therefore steep, sharp asperities are probably more sensitive to crushing than blunt ones. Of course sharp asperities in one direction may be broad flat asperities in another direction. The shape should therefore be considered in the direction of cutting (Figure 14).

An increase of the *spacing* may cause an increase of the load per asperity and therefore a reduction of the relief strength (Figure 15).

4.3.3 *Hardness of the mineral grains*

Mineral grains, which, under the condition of the cutting operation, are harder than the material of the cutting tool, are called abrasive. An abrasive mineral grain, which is in contact with the wear-flat of the rock cutting tool, may either break (crush) or penetrate the cutting tool or the rock. An increase of the hardness of a mineral grain

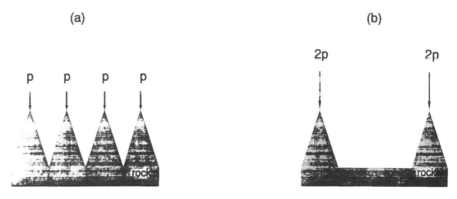

Figure 15 An increase of the spacing (from (a) to (b)) may cause an increase of the load per asperity

often causes an increase of the friability and a decrease of the toughness[9] of the grain (Malkin 1989). The hardness of mineral grains is hard to define since it is a material, geometry and test dependent property. Besides, temperature and stresses may affect hardness of minerals (and that of the cutting tool material) during rock cutting. Minerals, which are not considered abrasive to a cutting tool at room temperature for example, can be abrasive to the cutting tool at higher temperatures. In chapter 8 factors influencing hardness are discussed more extensively. An increase of hardness of the mineral grains in a rock may cause a change of abrasive capacity[10] of the rock.

4.4 WEAR MODE PREVALENCE

Three wear modes, described and illustrated in Figure 8, may occur, depending on the magnitude of the contact stresses relative to the strength of the rock relief. The cutting mode, scraping or cutting, is thought to affect the wear mode by influencing the cutting forces and the contact area between wear-flat and rock surface. The abrasive capacity of rock changes when the wear mode changes. This means that rock abrasivity is not an intrinsic rock property.

Kenny and Johnson (1976) distinguish three modes of rock-chisel interaction on the same basis. Zum Gahr (1987) distinguishes high and low stress abrasion. In high stress abrasion the particles are crushed during the process, while they do not fail in low stress abrasion.

In Figure 16 an overview of the parameters, discussed in the previous sections,

---

[9] Tough is described as strong and flexible; not brittle or liable to cut, break or tear easily (from the Longman dictionary of the english language, 2nd edition 1991).

[10] Cutting of abrasive rocks results only in abrasive wear if the cutting conditions favour abrasive wear and therefore abrasive rocks do not always cause abrasive wear; these rocks only have the capacity to cause abrasive wear under certain conditions.

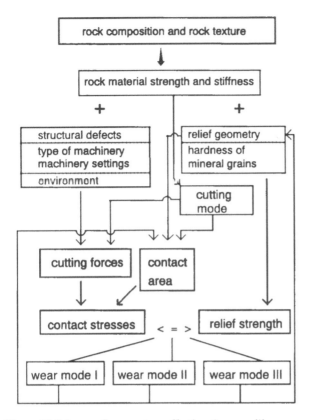

Figure 16 Scheme of parameters affecting the prevailing wear mode

and their influence on the relationship "contact stresses to relief strength" is given.

Wear mode I and II involve two-body wear. The abrasive action of the asperities in wear mode I is larger than in wear mode II because the asperities are intact in wear mode I, as opposed to the partly crushed asperities in wear mode II. In wear mode I temperatures are expected to be high due to high contact stresses at the asperity highs. Under these circumstances, the high temperatures can weaken the material of the cutting tool, by which minerals softer than the tool material at room temperature can become abrasive. In wear mode II also three-body abrasion may be involved. Wear mode II is thought to occur in the transition from scraping to cutting.

In wear mode III three-body wear is probably dominant. When the cutting forces are large relative to the rock relief strength the rock relief may be crushed, which causes a low abrasivity of the rock surface relief and three-body wear may then be the dominant abrasive wear mechanism. The temperatures in wear mode III are expected to be lower than in wear mode I and the cutting tool cuts the rock instead of scraping it.

The formation of rock chips of different sizes during the rock cutting process causes the feed of cutting tools into the rock to fluctuate (Figure 2). The fluctuation of the feed is possibly related to the variation of the cutting forces. Bisschop (1990), Davids and Adrichem (1990) measured the cutting forces during rock cutting

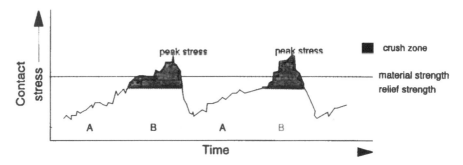

A: Time period at which the contact stresses are lower than the relief strength : low stress abrasion

B: Time period at which the contact stresses are higher than the relief strength : high stress abrasion

Figure 17 Contact stress in time. Variation is caused by the formation of rock chips of various sizes. Wear mode I occurs during low stress abrasion. Wear mode III occurs during high stress abrasion

experiments. The forces appeared not to be constant during the test. The forces increased to a maximum until a large rock chip broke out after which the forces fell back to a minimum (Figure 2). If the area of contact between chisel and rock is considered to be constant the contact stresses at the wear-flat vary as well during rock cutting.

The variable stresses result in different wear modes. Figure 17 illustrates how a rock relief responds to the stresses on the wear-flat in time. If the stresses acting on the relief asperities become larger than the strength of these asperities, the asperities get crushed.

Previous test results of wear experiments with a pin-on-disc type of test[11] already show a trend which support the proposed hypothesis of different wear modes occurring at different contact stresses (Deketh 1991).

It was found that with increasing pressures of the pin on the rock, wear increased at low pressures but decreased at higher pressures (Figure 18) This behaviour could be explained by a change of wear mode (two-body wear to three-body wear) when the relief strength is reached. Wear mode boundaries have arbitrarily been drawn in Figure 18. Other rock types show the same wear behaviour with an increase of the loading pressure. The position of the curves will be different for different rock types. Muro (1978) carried out a pin-on-disc type of test and found that with an increase of load on the "pin" rates of wear may change. Since the change of wear was accompanied by deepening of the groove in the rock, caused by sliding of the pin

---

[11] The modified pin-on-disc test

A flat-ended steel pin is pressed with a set force perpendicular on a rotating disc of rock. As time elapses the pin moves along the radius of the rotating disc of rock to slide over fresh rock surface. The velocity of the pin relative to the rock surface is constant during the test. The volume loss of the pin divided by the length of the travel path of the pin on the rock is a measure for the amount of wear ($W_{v/s}$ [m^3/m]).

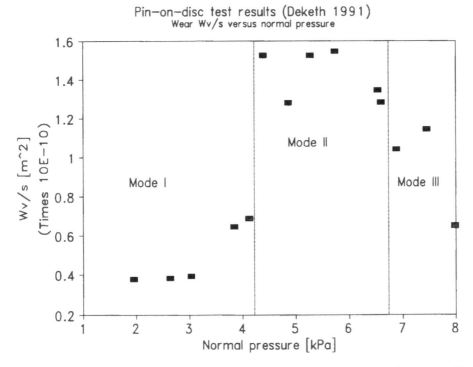

Figure 18 Pin-on-disc test results on an artificial lime-sandstone (Deketh 1991). Chisel wear $W_{v/s}$ [m³/m] versus normal pressure on the pin

over the rock surface, he assumed that the rate of wear changed as a result of the contact stresses exceeding the strength of the abrasive asperities.

CHAPTER 5

Rock cutting tests to study wear

5.1 INTRODUCTION

Tests, related to the rock cutting process, may be used as a means to clarify the causes of cutting tool wear, with respect to rock composition and texture. Rock cutting tests can be performed at various test-levels, the level becoming higher when the test resembles the real cutting process on a rock excavation site. Simple tests allow changing the variables more easily than complex tests. However, as laboratory tests become simpler, the experimental outcome may become less related to the mechanisms occurring during a real rock cutting project.

Verhoef (1994) distinguishes several testing categories for rock cutting tests, based on differences in the degree of simplification of in-situ processes during rock dredging. In Figure 19 the testing categories are shown. In Table 1 some authors, who carried out experiments related to rock cutting, are listed according to the test level of their experiments.

Most of these authors investigated the cutting process only, disregarding the wear process. Since wear processes and cutting processes are known to be closely interrelated, a review of investigations into rock cutting processes is worthwhile for the study of wear processes as well.

Wear due to rock cutting is a system dependent process (Uetz (1986). Therefore, simplification of the tests is allowed only, if the wear process does not deviate too much from the in-situ wear process. Model tests (category VI), such as the modified pin-on-disc test, which has been performed in the first phase of this work (Deketh 1989), are only remotely related to the processes occurring during rock practice because:

1.Loading conditions are different. The pin-on-disc test is force-controlled. The cutting process of a single chisel on a rock cutter head, like in rock dredging or rock trenching, is displacement-controlled.

2.Geometries of the pin and machined rock surface in the test differ from geometries of cutting tools and fresh rock surfaces in practice.

3.The interactions between pin and disc differ from interactions between cutting tools and rock in practice. No rock chips are formed with the pin-on-disc test.

Tests on rock types under similar loading conditions as in rock cutting practice (category V) are much more related to in-situ cutting processes. In the present

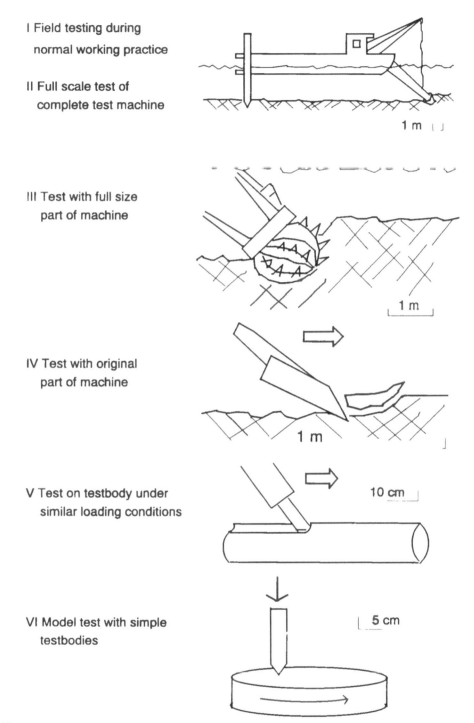

I Field testing during
  normal working practice

II Full scale test of
  complete test machine

1 m

III Test with full size
  part of machine

1 m

IV Test with original
  part of machine

1 m

V Test on testbody under
  similar loading conditions

10 cm

VI Model test with simple
  testbodies

5 cm

Figure 19 Testing categories related to rock cutting dredging. Simplification of the tests increases with an increase of the testing category number (after Verhoef 1994)

Table 1. A listing of some authors organised according to the test level of their rock cutting experiments. This listing of experiments is not related to a specific rock cutting method but covers various rock cutting disciplines like rock cutting dredging, tunnelling, trenching, ripping etc.

Test category Level	Authors
I & II	Lowe and Mc Queen (1990) ; Giezen (1993), Pells (1985) ; Verhoef (1993), Martin (1986) ; Abdullativ and Cruden (1983) ; MacGregor (1994)
III	Speight and Fowell (1984)
IV	Cools (1993) ; Kutter and Sanio (1982) ; Friedman and Ford (1983) ; Larson, Morell and Swanson (1987)
V	Dalziel and Davies (1964) ; Kenny and Johnson (1976) ; Whittaker and Swilzki (1973) ; Fowell ; Dubugnon and Barendsen ; Sman (1989) ; Bisschop (1991) ; Roxborough (1987)
VI	Schimazek and Knatz (1970, 1976) ; Paschen (1980) ; Verhoef, van den Bold and Vermeer (1990) ; Muro (1978) ; Deketh (1991) ; Mishra and Finnie (1979); Fowell (1970)

research a test has been designed, which belongs to a test level between tests of categories V and VI. The new test differs from a test of category V because small cutting depths can be reached and therefore the transition from scraping to cutting can be investigated. If weak rock types, with grain sizes which are relatively small compared to the feed, are tested, rock chips will be formed during cutting. If rock types with grain sizes of the same order of magnitude as the cutting depth are tested, single grains may break-out, but rock chips will not be formed. The saw-tooth behaviour of the cutting forces in time will not appear. However, it is considered to be important to look at the wear behaviour of rock cutting tools for various rock types at small cutting depths (feeds) (Kenny and Johnson 1976). With the new test, the processes occurring at transition of pin-on-disc tests (category VI) and cutting tests (category V) can be studied. As well as this, the test is displacement-controlled to simulate the process in practice. In practice the rock cutting machine as a whole is often force-controlled. However the individual cutting tools, which are rigidly fixed on the cutter head or boom of a rock cutting machine, are displacement controlled because of which the forces on the individual cutting tools vary.

5.2 THE SCRAPING TEST

The rock cutting test, designed for the present research, is suitable for study of the wear behaviour at the lower end of the range of feed of rock cutting tools as found, for example, in rock dredging or trenching. The rock-chisel interaction during this test is different from rock cutting at larger feed values. Large rock chips are not likely to form during the tests and therefore cutting forces do not relate to the formation of rock chips.

If weak rock types with small grain sizes are used, rock chips, as illustrated in

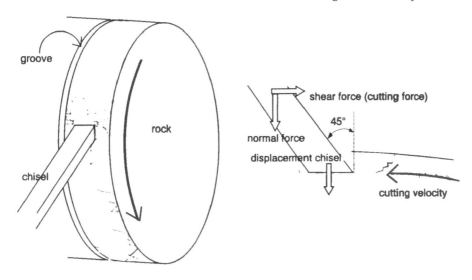

Figure 20 The principle of the scraping test experiments

Figure 8, may form in experiments at large feed values. At smaller feed values the chisel will scrape the rock surface, removing only very small fragments of rock. This process is called "scraping" and to distinguish this test from others, the test is called the "scraping test". The test equipment has been mounted on a special lathe[12], previously used for heavy duty metal cutting, because:

1.Easy and controllable variation of variables is possible. A major advantage of this lathe over other test rigs, such as a planer, is the possibility to reach high cutting velocities, sufficient to reach temperatures high enough to cause softening of the steel of the chisel.

2.It is practical. At the section Engineering Geology at Delft University of Technology a pin-on-disc test was mounted on this lathe. The scraping test could be set-up mounted on this lathe with a minimum of time and cost.

The experiments are based on rotating rock discs, penetrated by test chisels, which are described in chapter 8. In Figure 20 the principle of the scraping test is shown. To allow for cracks to propagate to free surfaces, grooves are sawn into the rock disc at a fixed spacing. The grooves simulate the result of the cutting of chisels travelling at both sides of the observed chisel. In Figure 21 the influence of the width of spacing on chisel interaction is shown. Friedman and Ford (1983) found that the specific energy (the energy needed to cut a unit volume of rock) decreased as soon as the spacing became so small that the chisels started to interact. For smaller cutting depths the interaction occurs at smaller spacings.

[12]Lathe: D.R.200 (machine nr. 17809) Artillerie Inrichtingen Zaandam Hembrug Nederland with a special motor drive (direct current motor (60 kW) with electronic control, feed is stepless variable).

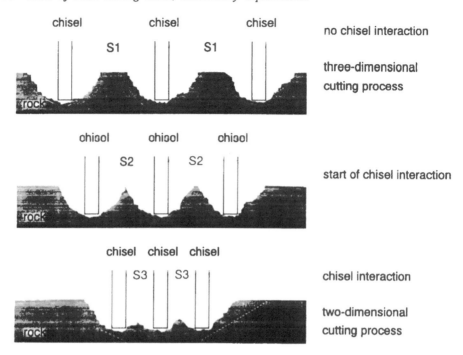

Figure 21 The effect of spacing on chisel interaction. S = Spacing. S1 > S2 > S3. The direction of cutting is perpendicular to the plane of the illustration

The lathe, used to carry the test arrangement, is shown in Figure 22. The feed[13] is kept constant for each test. The feed of the lathe can be varied from about 0 to 20 mm per revolution. The maximum value of feed in the experiments is about 2.5 mm per revolution; higher values of feed do not give repeatable results because the cutting path is too short. Most experiments are performed with a cutting path length of at least 25 m. When the chisel moves to the centre of the disc of rock, the angular velocity is automatically increased as to keep the cutting velocity constant. The r.p.m. (revolutions per minute) of the turning head of the lathe could be varied from about 20 to 5000 r.p.m. High r.p.m. values could however not be maintained at high feed values. Hence the cutting velocities tested range from 0.3 to 3 m/s. Cutting forces are measured in three perpendicular directions with a piezoelectric force transducer[14] (Figure 23). The signals from the piezoelectric force transducer are automatically recorded by a personal computer via charge amplifiers[15] and an A/D

---

[13] By the feed of machine tools is meant the distance the tool is moved at each revolution (or stroke) of the machine, i.e. at each revolution of the rock disc.

[14] Kistler 3 component dynamometer, type 9257 B. Maximum force in vertical direction (cutting force) is 7 kN at 145 mm centre distance. Maximum force in the other two perpendicular directions is 3.5 kN at that position.

[15] Kistler 3 channel charge amplifier type 5019A130.

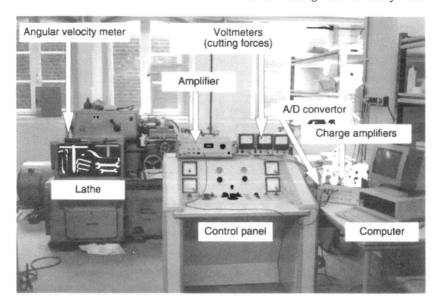

Figure 22 The test equipment

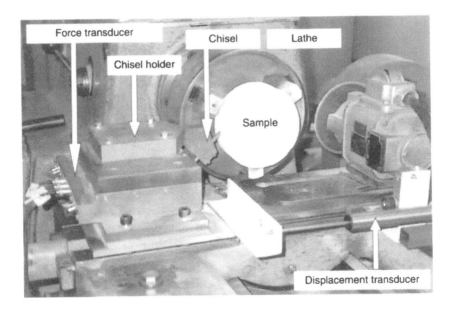

Figure 23 A close view on the test set-up

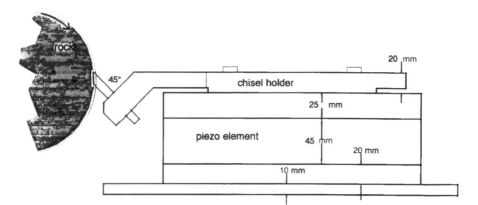

Figure 24 The specially designed chisel holder, which keeps the rake angle of the chisel at 45°

convertor. Also the angular velocity of the lathe[16], the radial displacement[17] of the support of the lathe and the temperature measurements by Chromel-Alumel thermocouples are automatically recorded by the personal computer via an amplifier and an A/D convertor. The data are processed and stored by the aid of the software package ASYST[18] on a personal computer. The chisel is kept in place and forces on the chisel are transduced to the force transducer by a specially designed chisel holder (Figure 24). The angular velocity is measured by an angular velocity meter and convertor. The radial displacement of the support of the lathe is measured with a displacement transducer. The displacement of the chisel tip into the rock per revolution of the lathe is the cutting depth and equals the feed of the support, minus the shortening of the chisel length due to wear per revolution. During the test the radial displacement of the lathe can be monitored on the LCD screen of the amplifier.

A frame with a plexiglass window to view the experiments is placed around the rotating rock disc penetrated by the chisel. This precaution is made to protect the operator from the dangers of a failing rock disc. Dust, produced during the cutting process, is removed by a vacuum cleaner during and after the experiment.

Before the test, the weight of the chisel is measured with an accuracy of 0.05 mg and calibrations of the force transducer and the thermocouple are carried out. The diameter of the rock disc is measured with a calliper (accuracy 0.05 mm) and fed into the personal computer together with the name of the sample.

---

[16] The angular velocity of the turning head of the lathe is measured at the turning axis with a r.p.m. remote measuring device and a signal processor, the ENI 5 and ENM3 both made by Hartmann & Braun AG.

[17] The (radial) displacement of the support of the lathe is measured with a long stroke displacement transducer.

[18] The data acquisition computer programme in ASYST has been written by W. Verwaal, Delft University of Technology, Faculty of Mining and Petroleum Engineering, Section Engineering Geology.

The wear-flat temperatures, the cutting forces, the cutting velocity and the feed are measured during each separate test. In each test run all variables are kept constant.

After the test, the weight of the chisel is measured and the wear-flat of the chisel and the rock surface can be photographed. Wear phenomena are described. Burrs are removed from the chisel before weighing. The forces recorded by the personal computer are averaged and stored together with the data on feed, cutting velocity, wear path length and test name on floppy disc.

The hardness of the tip of the chisel at various distances from the wear-flat can be measured. The diameter of the rock disc after testing is measured.

CHAPTER 6

Rock properties with respect to wear

6.1 INTRODUCTION

This study aims to get a better understanding of the effects of rock properties on the rate and type of wear in the transition from scraping to cutting, by performing small scale rock cutting laboratory experiments[19]. Also the cuttability (the facility of intact rocks to be excavated by cutting tools) of the different types of rock is considered, since this property is thought to be related to the wear capacity of rock. Various authors investigated which rock properties affect wear and derived wear index values from experimental results by combining various rock property parameters. Although these index values closely relate to the results of their experiments, the correlation with actual tool wear in rock cutting projects is often very poor or limited to specific machinery or conditions. However, the index values show which rock properties may be of significance to the rate and type of tool wear under specific conditions. Some index values are reviewed below.

6.1.1 *Rock property based rock cutting wear index values*

Schimazek's wear index value "F" resulted from wear experiments, using a pin-on-disc test, on Carboniferous sedimentary rocks (Schimazek and Knatz 1970, 1976) . The F-value is linearly related to the wear rate of the pin.

$$F = 10*(HQ_{eq})*\Phi*BTS \qquad [N/m]$$

HQ_{eq} = *Equivalent quartz volume percentage* [%]
Φ = *average grain size*   [mm]
BTS = *Brazilian tensile strength* [MPa]

(1)

---

[19] In this study rock is considered to be a cemented or crystalline aggregate of mineral grains or crystals (as well natural as artificial). The Unconfined Compressive Strength of rock is at least 1 MPa.

Ewendt (1989) modified the F-value of Schimazek as a result of correlation of the index value with disc cutter tests. The modified F-value, F_{mod}, is linearly related to the wear rate of the disc cutter.

$$F_{mod} = HQ_{eq} * Is_{50} * \sqrt{\Phi > 1mm} \qquad [N/mm^{1.5}]$$

HQ_{eq} = *Equivalent quartz volume percentage* [%]   (2)

Is_{50} = *Point load strength* $[\dfrac{N}{mm^2}]$

Φ = *average grain size*   [mm]

McFeat-Smith and Fowell (1977) derived a wear index value, CW (Cutting Wear), from experiments with the core cuttability test, as described by Roxborough (1987). The wear index CW is linearly related to the wear rate of the used tool.

$$CW = 0.55 + 4.25SH^3 * 10^{-5} - 1.88SH^2 * 10^{-3}$$
$$+ 1.98CC^3 * 10^{-3} + 1.2QC^3 * 10^{-6} \quad [mg/m]$$

                                                                                (3)

SH = *Shore Hardness*
CC = *Cementation Coefficient*
QC = *Quartz Content*

Rock property parameters, often combined into a formula yielding an index value like in the Equations 1,2 and 3, are often used to describe rock behaviour in rock cutting processes. The index values correlate often only closely to the results of the test, which forms the basis of the index values, and not to rock cutting processes in general because:
1. Index values only based on rock properties may lack important non-rock property factors[20], which are essential for a good correlation of index values with actual rock behaviour in different rock cutting processes.
2. Some rock parameters used in the index values are not inherent properties of the rock itself, but depend on the circumstances of testing or cutting.
In the following section some attention is paid to the second remark.

6.1.2 *Some considerations on the non-intrinsic rock property parameters*

The parameters, which describe the rock may be divided into the following two groups:
1. Physical rock property parameters such as grain size, density and porosity. These parameters describe intrinsic rock properties. The parameters are inherent only to the rock itself.
2. Mechanical rock properties like strength and hardness. These rock properties are

---

[20] Some of these factors are: geometry and material of the cutting tool, operating conditions like feed and cutting velocity of the cutting tool, type of rock cutting machine.

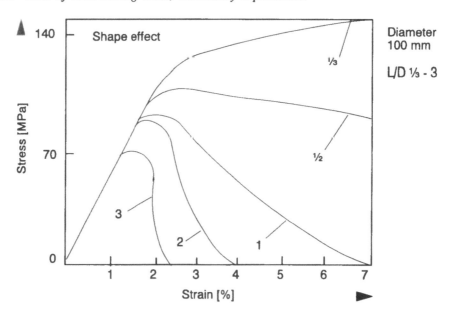

Figure 25 Influence of core shape expressed in terms of length/diameter (L/D) ratio for a core with a diameter of 100 mm for Georgia Cherokee Marble (after Hudson et al. 1971 in Farmer 1983)

parameters, which are influenced by the method of testing.

Mechanical rock properties may therefore only be used to describe the behaviour of rock if the nature of testing is comparable or the same as the nature of the process to which the rock is subjected. Two parameters, rock material strength and hardness of the minerals, are often used to describe behaviour of rock and its components in rock cutting processes. Since these parameters are considered to be of importance to wear of rock cutting tools by many authors, some attention is paid to them.

6.1.2.1 *Rock material strength*

Many different tests exist to determine the strength of rock. The difference made between tensile strength (ASTM D3967-81) and compressive strength (ASTM D2938-80) already shows that there is no such rock property as, simply, "strength". Apart from the test conditions, such as strain rate, temperature and confining pressure, also the properties of the test apparatus, such as stiffness, and the dimensions of the test piece influence the final compressive and tensile strength value. For instance, a rock cylinder tested in uniaxially unconfined compression will behave differently at a length/diameter ratio of 0.3 than at a length/diameter ratio of 3 (Figure 25, Farmer 1983). Loading conditions in standard rock strength tests differ from those in rock cutting. An increase of cutting speed during rock cutting for example gives rise to higher strain rates in the loaded rock. Rock will appear stronger at higher strain rates (Reinhardt 1985). In standard strength testing, however, strain rate is not adapted to the strain rates of the various rock cutting processes taking place in practice.

Strength and failure mechanisms of rock depend on the confining pressure. A higher confining pressure is related to a higher strength of the rock. Also the forces needed to penetrate a confined rock with a pick are higher at higher confining

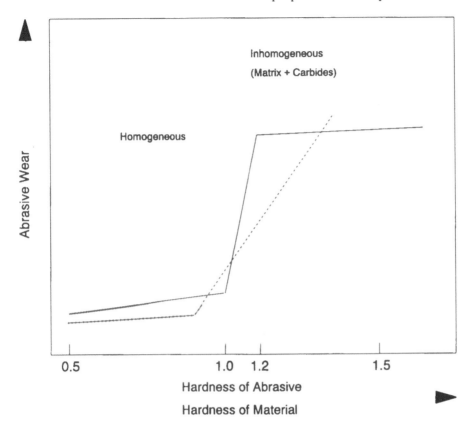

Figure 26 Abrasive wear as a function of the ratio of hardness of the abrasive to hardness of the abraded material (after Zum Gahr 1987)

pressures (Larson, Morrell and Mades 1987). At high confining pressures rock, which under low confining pressure conditions behaves in a brittle way, will behave more ductilely. Whether a rock behaves in a brittle or a ductile manner is thought to be of importance to rock cutting processes (Cools 1990).

The examples described of how some factors affect the strength of rock in rock cutting or strength testing show the limitations of applying the results of a standard strength test (such as UCS or BTS) to describe the cutting characteristics of rock with respect to different rock cutting machinery under different operating conditions and stress conditions in the rock.

6.1.2.2 *Hardness of minerals*
Hardness is often regarded as an important property of steel as well as rock to describe their behaviour in rock cutting processes, as far as abrasive wear is concerned. Both rock and steel are often inhomogeneous on the scale of testing and may consist of several components of different hardness. Thus "the hardness of rock or steel" is a combined value of the hardness values of their components, some of them influencing the hardness more than others. Carbides in steel for example have

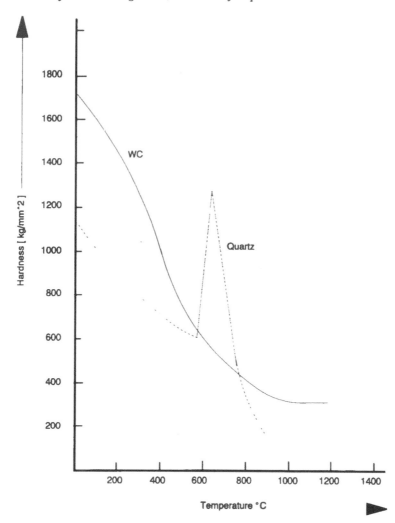

Figure 27 Hardness of tungsten carbide (WC) and quartz as a function of the temperature (after Osburn 1969)

a significant effect on the resistance to wear of steel cutting tools, but do not influence the overall hardness (Vickers) since they are too small to be significant for the Vickers hardness determination.

Studies on wear showed that abrasive wear is a function of the relative hardness of the materials involved (Uetz 1986). In Figure 26 shows the relationship between the ratio hardness of abrasive / hardness of material and abrasive wear. It has been established that one material will scratch another provided that there is a large enough difference between their respective indentation hardnesses ($> \approx 20\%$).

However, the relative hardness of steel and rock minerals is insufficient to describe their behaviour in a wear system. This is partly due to the different natures of rock

Table 2. Hardness of some minerals according to three different test methods (after Verhoef 1990 and West 1981)

Mineral	Hardness			
	Vickers	Mohs	Rosiwal	
Talc	20	1	0.3	
Halite or Gypsum	50	2	1.25-2	softer than fingernail
Calcite	125	3	4.5	harder than fingernail
Fluorite	130	4	5	
Apatite	550	5	6.5-8	equivalent to copper coin
Feldspar	750	6	37	equivalent to windowglass
Quartz	1000	7	120	equivalent to penknife
Topaz	1850	8	175	equivalent to hard file
Corundum	2300	9	1000	
Diamond	10060	10	140000	

and steel and their different mechanical response in hardness testing and wear systems.

Besides this, factors such as temperature, affecting the wear resistance of steel in wear systems, are not incorporated in standard hardness testing. Moreover, high temperatures, which may develop at the interface of rock and steel in cutting operation, affect the hardness of steel in a different way than the hardness of rock minerals. Although tungsten carbide, often used for cutting tools, is harder than quartz at a temperature of 20 °C, quartz is harder than tungsten carbide at temperatures between 600 °C and 800 °C (Osburn 1969) (Figure 27). The effect of high temperatures in wear systems is not included in standard hardness tests.

The results of hardness tests depend not only on the nature of the material but also on the nature of the test itself. Tests, like the Vickers or Brinell hardness tests, which take the indentation diameter of an imprint of a sphere or pyramid, pressed with a prescribed load and period of time into the test body, are, as a measure of hardness, completely different from tests like Rosiwal's hardness test, which uses the loss of material after grinding the test body with abrasive powder as a measure of hardness or the Mohs hardness test, which uses the ability of one mineral to scratch another as a measure of hardness. Diamond, which cannot be scratched by any other mineral and is thus the hardest, has a hardness value of 10. Talc, which cannot scratch any other mineral, has a value of 1. In Table 2 the hardnesses of some minerals according to three test methods of different nature are listed (Verhoef et al. 1990). It can be seen that the hardness scales are not linearly related.

Despite the many drawbacks of using non intrinsic rock properties, like strength and hardness, to describe rock behaviour in rock cutting processes because of lack of better alternatives they are still considered to be descriptors of rock behaviour in cutting processes.

CHAPTER 7

Experiments

7.1 INTRODUCTION

During cutting of rock by a rotary cutting machine like a cutter suction dredger or a rock cutting trencher the feed (the displacement rate of a chisel into the rock) gradually increases when the chisel proceeds along its arc-shaped cutting path. Besides, the momentary feed varies with the formation of rock chips of different sizes. The general increase and the momentary variation of the feed of the chisel causes a gradual increase and a momentary variation of stresses in the contact between rock and chisel. As a result the wear modes may change from one to another during the cutting process (chapter 4). Experiments have been designed to investigate the change and type of wear mode by variation of the displacement rate of a chisel into different types of rock with two types of chisel steel.

The objectives of the experiments were:
- to show that an increase of the feed of the chisel into the rock can lead to a change of mode (type and rate) of wear.
- to show that this behaviour is caused by the occurrence of different wear modes as described in the wear mode hypothesis.
- to investigate the influences of some rock material properties on the different wear modes.
- to investigate the effects of the cutting velocity on the rate and type of wear.
- to investigate the influences of some rock properties on the type and rate of wear at different cutting velocities.

Information on the influence of rock properties on the cuttability of rock, which is thought to be related to the wear capacity of rock, is considered to be a secondary result of these experiments. Rates of wear of the test chisel were compared to the energy needed to cut the rock.

7.2 SCRAPING TEST EXPERIMENTS

Initial experiments have been carried out to modify the boundary conditions of the test and to determine the repeatability of the test. The shape of the chisel, for example, has been designed in such a way, that the initial shape of the chisel does

Table 3. Parameters describing wear and cutting processes

Parameter	Explanation	Unit
Feed	radial displacement of the support of the lathe per revolution of the turning head of the lathe	mm/revolution
Cutting depth	thickness of the rock-cut per revolution of the turning head of the lathe	mm/revolution
Cutting velocity	velocity of the chisel relative to the rock at the chisel-rock contact in the direction of cut	m/s
Cutting force (Fz)	force acting on the chisel parallel to direction of cut (also named shear force or tangential force)	N
Normal force (Fy)	force acting on the chisel perpendicular to the rock surface	N
Rate of wear	difference in weight of the chisel before and after an experiment divided by the length of cut	g/m
Specific wear (SPW)	difference in weight of the chisel before and after an experiment divided by the volume of rock cut	kg/m^3
Cutting mode ratio	ratio of cutting depth to feed	--
Specific energy (SPE)	cutting force multiplied by length of cut and divided by the volume of rock cut	MJ/m^3

not change much as the wear-flat wears away during testing. For a changing shape of the chisel during testing may affect the wear process and since the effect of chisel shape on the wear process is no subject of investigation, this should be avoided.

Before a series of tests is run with varying factors, the repeatability of the test should be known. No proper interpretation of test results can be made without knowing the repeatability[21] of the test.

After tuning the test to the results of the initial experiments, experiments were designed to meet the proposed research objectives.

The recorded or measured test parameters (chapter 5) were reworked to parameters which describe the wear and cutting processes.

The rate of wear can be expressed as the weight loss of a chisel per unit volume of rock cut (the specific wear SPW) or per meter of cutting distance (rate of wear). The SPW value shows the efficiency of the cutting process with respect to wear. The rate of wear indicates how fast a chisel wears away independent of the amount of cut rock material.

The total volume of rock cut in an experiment can be calculated with the help of

---

[21] Repeatability $r = 1.96 \times \sqrt{2} \times \sigma$ (σ = standard deviation) (BS 812:1975). Repeatability is the quantitative expression of the random error associated with a single test operator obtaining successive results on identical material, with the same equipment an constant operating conditions. It is the difference between two single test results, which could be expected in only one case in twenty, assuming normal and correct operation of the test.

Table 4. Parameters varied in the experiments

Group	Parameter
Artificial rock	Mineral content
	Volume % abrasive minerals
	Strength
	Grainsize of abrasive minerals
	Angularity of abrasive minerals
Natural rock	5 types of sandstone
	2 types of limestone
Steel	Wedge steel
	Dredge steel
Machine/test settings	Feed
	Cutting velocity

the diameter of the rock disc before and after the test. The total volume of cut rock divided by the number of revolutions of the disc is a measure for the depth of cut. The depth of cut will be compared to the radial displacement of the support of the lathe (feed). The ratio of the depth of cut to the feed (cutting mode ratio) reveals whether the chisel cuts or scrapes the rock. A value close to 0 indicates scraping and a value near 1 indicates cutting.

The energy to cut a unit volume of rock, specific for the lathe test set-up (specific energy, SPE), is a measure for the cutting efficiency of the process and the cuttability of the rock. The specific energy is obtained by multiplying the cutting force by the cutting distance and dividing this by the volume of rock cut.

In Table 3 the parameters, which describe the wear and cutting process, are listed.

The investigations were designed to study the influences of individual rock properties on the type and rate of wear. Artificial rocks (mortars) of various compositions were used (chapter 9), which allowed easy, controllable and independent variation of rock property parameters. Most of these mortars were composed mainly of Portland B cement and quartz sand, since according to many authors quartz is considered to be of prime importance to the abrasivity of rock. Experiments have been carried out with chisels made out of two types of steel (chapter 8). Dredge steel (SRO 57N) test chisels were made out of the (new) chisels of rock cutting dredgers. Preparation of these chisels was costly and time consuming, but the use of these chisels in the experiments was valuable because wear processes are known to depend very much upon the properties of the tool materials involved. Wedge steel (Fe 60K) test chisels could be prepared more easily and therefore allowed more attention to be paid to the study of the effects of the various rock properties on the wear process. The advantage of using both types of steel, was that the relative ease of preparing test chisels of wedge steel made the performance of many tests possible, while the results, obtained with dredger steel, were more realistic, i.e. more relevant for practice.

The feed and cutting velocity, being test settings, have been varied in different test runs as well as several rock properties.

During a single test run all variables were kept constant.

Microscopic examination of the wear-flat of the chisel and the rock surface after a test was performed, because this might give evidence of the presence of different wear modes for different machine settings and different rock compositions.

Natural rock types were used to verify the experiments on artificial rock. The rock types used in the experiments are described in chapter 9. Five types of sandstone and two types of limestone have been used. Sandstones resemble the artificial rock most and were therefore expected to react in the same way with respect to the prevailing wear and cutting processes. Limestones were also used since they do not resemble the structure of the artificial rock and may therefore show a different response with respect to the wear and cutting processes. This might show the limitations of the experimental results found on artificial rock.

In Table 4 a comprehensive overview is given of the parameters varied in the experiments.

CHAPTER 8

Chisels used in the experiments

8.1 INTRODUCTION

The type of cutting tool influences the type of rock cutting mechanism, the energy required to excavate a unit volume of rock (specific energy), the resistance to wear and failure etc. (Phillips and Roxborough 1981; Dalziel and Davies 1964). Cutting tools can be distinguished on the basis of their shape or of the material they are made of. The differences between cutting tools may be reflected in their names. Bray (1979), for example, distinguishes three different types of cutting tools for dredging; the pick point, the flared point and the chisel point. In trenching the cutting tools are called bits (Giezen 1994) and in tunnelling and mining the cutting tools mounted on the rock excavation machinery are often called picks (Phillips and Roxborough 1979). This chapter will treat the chisel[22] types, used in the experiments of this research.

8.2 CHISEL MATERIAL

Materials used for rock cutting tools are mostly steel and tungsten carbide. The pick-points used for rock dredging are often made of steel, whereas the bits of rock cutting trenchers are made of steel with a tungsten carbide insert. Tungsten carbide is much harder than steel and has therefore a higher threshold to abrasive wear; minerals, of a Vickers hardness between 600 and 1200 Vickers, are abrasive to steel, used for rock cutting tools, but not to tungsten carbide. A disadvantage of tungsten carbide is its higher brittleness and its sensitivity to failure under impact loading conditions. Moreover, if rocks containing minerals harder than tungsten carbide are cut, abrasive wear takes place more severely than in case of steel cutting tools. The harder tungsten carbide suffers from more abrasive wear than steel because of a different failure mechanism; tungsten carbide fails by microcracking, whereas steel fails by microploughing and microcutting (chapter 3).

The steel supporting or holding the tungsten carbide insert (e.g. bits of rock cutting trenchers) usually wears away faster than the tungsten carbide insert, resulting in an

---

[22] According to the Longman English dictionary a chisel is a metal tool with a cutting edge at the end of a blade.

Table 5. Composition and properties of steel SRO 57N (dredge steel) and Fe60K (wedge steel). The hardening and tempering scheme of SRO 57N is kept secret by the steel manufacturer (Bofors), the chisels are made from cutter teeth delivered by the shipyard Stapel

Fe60K		SRO 57N	
Composition %	Properties	Composition %	Properties
0.4 C		0.29 C	yield stress 1350
0.25 Si		1.6 Si	failure stress 1600
0.65 Mn		1.2 Mn	strain limit 5 %
< 0.05 P		0.5 Ni	incision 10 %
< 0.05 S		0.2 Mo	notch value J 25
	hardness 300 VH		hardness 614 VH

insert not supported by steel anymore. Therefore, the insert will fall out of the steel holder leaving the steel unprotected to further wear.

Tungsten carbide is more expensive than steel and the use of tungsten carbide may therefore be not as economic as the use of the more wear-sensitive steel.

Alloy elements like Co, W, Mo, Si and Cr are often used in steel to improve its properties as rock cutting tool. W and Si can form carbides, which improve the wear resistance of steel. Co gives steel the ability to respond to impact loadings. Steel containing much Mn like Hadfieldsteel (13% Mn and 1.3% C) hardens under impact loading and is therefore often used for tools, such as jaw crushers, rock drilling bits, dredging tools, which are subjected to wear due to large impact loadings (Korevaar and Pennekamp 1984). Cr and Mo have a positive effect on the hardening of steel.

The wear-behaviour of the different steel types or that of tungsten carbide is interesting but is beyond the scope of this research. Since wear of steel seems to be more critical than of tungsten carbide in dredging, the study of the wear of steel is emphasized. Two types of steel have been tested in this research: SRO 57 N and Fe60K. Their composition and hardness is listed in Table 5.

SRO 57N cast steel is used for cutter teeth of cutter suction dredgers. The steel resembles most other cast steel types used for rock cutting chisels. This type of steel has a high resistance to fatigue and a high sensitivity to decarbonization (Korevaar and Pennekamp 1984).

Fe60K steel is used for wedges and will be used for comparison of wear phenomena with steel SRO 57N. Fe60K has been chosen rather arbitrarily. It was readily available, could be easily machined and was thought to wear away in measurable quantities.

8.3 CHISEL SHAPE

The chisel shape influences the prevailing cutting mechanism (Figure 10), and thus

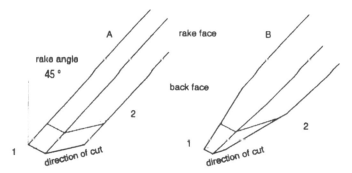

Figure 28 The wear-flat of the test chisel (A) remains constant and the wear-flat of the pick-point chisel (B) increases when wear progresses (from 1 to 2)

may affect the temperatures at the rock-steel contact as well as the cutting forces necessary to cut the rock. During the rock cutting process the wear-flat, attack face and sides of a chisel wear away. This may alter the initial shape and thus the cutting forces (van der Sman 1988), the initial cutting/wear mode and the temperatures at the wear-flat may change.

Results of the experiments with the scraping test can therefore only be compared if the initial chisel shape persists. During the scraping test wear will mainly occur at the wear-flat. Therefore a suitable chisel shape has been chosen to keep the wear-flat area and the shape as constant as possible as the chisel wears away during testing.

To meet this condition the rake and back face of the chisel were made parallel as well as the side faces. As the wear-flat wears away, the wear-flat area remains constant. In Figure 28 the test chisel used in the experiments is shown (A) and compared to a pick point type chisel (B). When the chisels wear, from stage 1 to stage 2, the wear-flat of the test chisel remains constant, whereas the wear-flat of the pick-point type chisel increases. The rake face has an angle of 45 degrees with the direction of cut. This value lies within the range of rake angles used in practice. The wear-flat has a length of 7 mm and a width of 5 mm.

8.4 THERMOCOUPLE

To measure temperatures a thermocouple can be installed near the wear-flat of the chisel. H.P.H. van Rossen (1988) tested various positioning alternatives of thermocouples.

Thermocouples, positioned at the wear-flat of the chisel, are susceptible to damage and may not measure the temperatures at the wear-flat, but measure the average temperatures of the thermocouple joint. Since this joint is not in full contact with the wear-flat, temperatures, measured with the thermocouple, may be lower than the temperatures at the wear-flat. Locally, at the contacts of rock asperities and the chisel, much higher temperatures of very short duration, the socalled flash-temperatures, may occur, but are not detected by the thermocouple.

Thermocouples positioned at some distance from the wear-flat, and thus from the heat-source, are less susceptible to damage, but may only indicate the height of average temperatures at the wear-flat.

Absolute temperatures at the wear-flat of a chisel during cutting can approximately be assessed by microscopic examination of the texture of the steel and by micro-Vickers hardness measurements of a chisel section after an experiment. As the steel texture and hardness is also affected by high contact stresses, precise deduction of the temperatures at the wear-flat during a cutting experiment is impossible.

CHAPTER 9

Properties of rock types used in the experiments

Changing one property at a time is necessary if data are to be obtained for evaluating the effects of each property on the total wear process. Since natural rock types cannot meet this requirement, artificial rock types were used to determine the effects of changing one property at a time. Mortar types of different compositions seemed to be suitable. Differences in material strength by changing composition can be overcome by the property of cement to increase in strength as time proceeds or by the addition of small quantities of silica fume or by variation of the water/cement ratio. The composition and properties of the used mortars can be found in table 8.

For initial experiments artificial lime-sandstones have been used. These rock types were also used in rock cutting wear investigations by Deketh (1991) and Bisschop (1991). Some properties of these rock types are shown in Table 6.

Also natural rock types have been tested to compare the expected rate and type of wear with the results of experiments on the mortars. Five types of sandstone and two types of limestone have been used: The natural rock types were superficially homogeneous and isotropic in the test specimen and were expected to be suitable to carry out a number of experiments with a high repeatability. The petrographic description of the natural rock types is given in appendix II. The properties of the natural rock types can be found in Table 9.

The sandstones contain quartz grains, which are able to scratch the steel of the chisel due to the higher hardness of quartz. This may lead to abrasion (grooving wear).

The limestones are composed of calcium carbonate, which is softer than the steel used, at least at room temperature or/and at low contact stresses. At low temperatures or/and low contact stresses wear will probably take place by sliding wear (plastic deformation, ploughing, adhesive wear) At higher temperatures and/or higher contact stresses, abrasive and adhesive wear may occur. In chapter 3 these rock cutting wear mechanisms and factors which control them are discussed more extensively.

9.1 ROCK MATERIAL PROPERTIES

Many different parameters, retrieved from a large range of tests, can be used to

Table 6. Some rock properties of rock types, artificial calcarenites, used for initial experiments

rock name	quartz volume percentage (%)	grain size (mm)	UCS (MPa)	BTS (MPa)	density (Mg/m³)	porosity (%)
A	53	0.57	16	1.5	1.9	28
B	46	0.20	9	1.4	1.6	39
H	51	0.42	15	1.5	1.7	34

describe the material properties of rock related to rock cutting wear processes (chapter 6).

A selection of these parameters and their accompanying tests is made using the following selection criteria:

1. The parameters should have proven to be of significance to rock cutting and to tool wear processes in former research or in practice.
2. The test, which yields such a parameter, should be reliable and widely accepted as a standard means to determine a specific rock property.
3. The test should be simple and easy to perform to allow for application in practice and multiple testing.

The list of parameters and tests presented in Table 7 resulted. Tensile and compressive strength of mortar samples were determined after ± 28 days of curing time in a room of 95% humidity. The tensile strength was determined with the Brazilian Tensile Strength (BTS) test, the compressive strength with the Unconfined Compressive Strength (UCS) test and the Elasticity (E) Modulus was measured at 50 % of the UCS. Before testing, the samples were oven dried (± 90 °C) during ± 24 hours. Specific gravity and dry density were determined and from these parameters the porosity was calculated. These rock property parameters have also been determined for Felser sandstone, Bentheimer sandstone, Elb sandstone type C and D and sandstone T. In Tables 8 and 9 the rock property parameters of the different types of mortar and those of the natural rock types are shown. The standard deviation of the strength parameters shown in these tables all have a standard deviation lower than 15 % of the presented value. All values are averages of more than 5 identical tests (in case of the UCS more than 3 identical tests). The standard deviation of the point counting (500 counts) was less than 2.5 %. The specific gravity was determined with a standard deviation smaller than 0.15 g.

9.1.1 *Interpretation of the measured and derived rock properties*

The mortar compositions of Table 6 have been used to see whether (linear) relationships exist between the parameters, which are considered to describe the behaviour of rock during cutting. If a (linear) relationship exists between two parameters it is sufficient to investigate the relation of only one of them with the wear and cutting processes. In Figure 29 a correlation matrix of the parameters considered is shown. The Rock Toughness Index (RTI = $(UCS^2/(2*E\text{-modulus})*100$ expressed in MPa) is also considered since it showed a good relationship with the

Table 7. Rock parameters and accompanying tests

Mechanical Properties	Parameters	Test standard or method
Strength	Unconfined Compressive Strength (UCS)	ASTM D2938-80
		ISRM
	Point load strength (Is50)	ASTM D3967-81
	Brazilian Tensile Strength (BTS)	
Elasticity	Elasticity Modulus (E)	ASTM D3148-80
Mineral hardness	Vickers Hardness Number(VHN)	
	Mohs hardness (Mohs)	
	Rosiwal hardness	
Physical Properties		
Dry density	Dry density	ASTM C97-83
Specific gravity	Specific gravity	BS 1377:1975, Test 6(B)
Grain size	Grain size (Φ)	Thin section analysis
Grain Shape	Powers Roundness (scale 0-1)	Powers (1953)
Mineral composition	Volume percentages	Point counting

performance of rock cutting trenchers (Giezen 1993). A good, moderate and bad relationship, based on visual interpretation of scatter plots is marked respectively by a +, +/- and a - sign. Below the signs the correlation coefficients are shown which are a measure of how well two parameters are linearly related.

In Figure 30 some scatter plots of parameters which showed a good or moderate relationship are shown.

BTS and Is50 show a good linear relationship and probably describe the same rock property, the tensile strength. A linear relationship between UCS and BTS or Is50 is found if only samples with approximately the same volume percentage of abrasive minerals[23] are included.

The ratio of compressive strength and tensile strength may indicate a brittle or ductile behaviour of the rock (Gehring 1987); a low ratio indicating ductile and a high ratio indicating brittle behaviour.

The ratios of UCS over Is50 (UCS/Is50) and of UCS over BTS (UCS/BTS) decrease if the volume percentage of abrasive minerals increases (volume percentages are adjusted for porosity) (see Figure 31). Young's E-modulus increases generally with an increase of tensile or compressive strength. The Rock Toughness Index (RTI) relates well to the volume percentage of abrasive minerals, porosity, UCS and the ratio UCS/BTS.

The use of independent parameters to study the influence of rock properties on rate and type of wear is favoured since a set of independent parameters is considered to

---

[23] The abrasive mineral in the mortars is quartz except for mc 16 and mc 17. The abrasive mineral for mc 16 and mc 17 is respectively glass and potash-feldspar.

Table 8. Composition, mechanical and physical properties and vibration time of 17 mortar compositions. Cement type is Portland B. The mortars are casted into 3 cubes of 8000 cm³. After casting the mortars, the cubes are vibrated on a vibration table. Strength values (UCS, BTS, Is50 and E-modulus) of the mortars are related to strength values of dry samples after a curing period of 28 days in a room with 95% humidity. UCS is Unconfined Compressive strength. BTS is Brazilian Tensile Strength. Is50 is Point Load Strength. ls50 is Point Load Strength (diametral). SF = Silica Fume. lg,n = ligno gluconate, naphtalene. m = melment. Volume percentages of abrasive minerals within the brackets are corrected for the porosity. The type of abrasive minerals is in most mortars quartz, except in mc 16 (silica-glass) and in mc 17 (potash-feldspar)

mortar composition	W/C factor	additives	vibration time (s)	vol % abrasive minerals	grain-size minerals (mm)	grain shape minerals	UCS (MPa)	BTS (MPa)	Is50 (Mpa)	E-modulus (GPa)	dry density (Mg/m³)	Porosity (%)
mc 1	0.4		120	65 (58)	1-2	rounded	30	2.3	1.6	19	2.2	10
mc 2	0.6		30	65 (53)	1-2	rounded	18	1.7	1.3	11	2.0	18
mc 3	0.2	SF,lg,n	120	65 (58)	1-2	rounded	64	4.7	3.6	30	2.2	10
mc 4	0.4	m	120	65 (58)	0.5-1	rounded	24	2.4	1.4	21	2.1	11
mc 5	0.4	m	120	65 (59)	0.25-0.5	rounded	26	2.4	1.2	9	2.2	9
mc 6	0.6	SF,m	180	65 (53)	0.125-0.25	rounded	23	1.9	1.4	10	1.9	19
mc 7	0.3	SF,lg,n	60	65 (61)	0.5-1	rounded	72	5.3	4.1	30	2.3	7
mc 8	0.4	SF,lg,n	30	65 (50)	0.5-1	rounded	44	2.7	3.1	18	1.8	24
mc 9	0.4		30	50 (46)	1-2	rounded	37	2.6	2.5	19	2.1	9
mc 10	0.6	SF,lg,n	60	65 (50)	1-2	rounded	17	1.0	1.0	8	1.9	23
mc 11	0.4		0	35 (33)	1-2	rounded	58	3.0	2.1	13	1.9	6
mc 12	0.4		0	20 (19)	1-2	rounded	67	2.7	2.0	11	1.9	6
mc 13	0.4		0	20 (19)	0.25-0.5	rounded	66	3.3	2.0	11	1.9	6
mc 14	0.4		0	0			65	2.4	1.3	10	1.9	3.2
mc 15	0.4		240	65 (48)	1-2	angular	16	2.2	2.0	6	1.8	26
mc 16	0.4		240	65 (50)	1-2	angular	9	1.1	0.9	5	1.6	23
mc 17	0.4		240	65 (52)	1-2	angular	14	2.5	1.5	8	1.8	20

Table 9. Properties of natural rock types, used for experiments. UCS is Unconfined Compressive strength. BTS is Brazilian Tensile Strength. Is50 is Point Load Strength (diametral). All rock property values are average values of at least 3 specimens. Porosity has been calculated from specific gravity and dry density measurements. Porosity values within the brackets are values of other authors (Hettema et al. 1991 measured the porosity of felser, de Roos 1991 measured the porosity of bentheimer and van der Meer (1979) determined by thin section analysis the porosity of Euville and Sirieul). These values give an impression of possible differences between measurement methods. Volume percentage and grain size and shape of abrasive minerals have been determined from thin sections by point counting and disregard the porosity. The values of the volume percentage of abrasive minerals within the brackets take the porosity into account

rock name	vol % abrasive minerals	grain size sand (mm)	grain shape	UCS (MPa)	BTS (MPa)	Is50 (MPa)	E-modulus (GPa)	dry density (Mg/m³)	porosity (%)
Felser	67 (54)	0.18	angular (0.20)	30	2.6	1.7	7	2.3	21 (20)
Bentheimer	93 (71)	0.11	subangular (0.35)	59	3.4	2.9	19	2.0	24 (24)
Elb-sandstone C	92 (69)	0.12	subangular (0.30)	35	2.9	1.8	14	2.0	25
Elb-sandstone D	91 (72)	0.35	subangular (0.30)	57	3.7	2.8	22	2.1	21 (20)
sandstone T	86 (69)	0.34	(sub)angular (0.25)	54	3.5	2.4	21	2.2	20
Sirieul	5 (3)	0.16	subangular (0.30)	9	1.1	0.9	6	1.8	36 (30)
Euville	0			25	1.9	1.5	23	2.2	20 (20)

	UCS	BTS	Is50	Grain size	Grain shape	Dry density	Porosity	E-modulus	RTI	Vol.% abr.min	UCS/BTS
UCS	1	+/- .77	+/- .64	- -.16	- .54	- .09	+/- -.76	+/- .53	+ .84	- -.53	+/- .75
BTS		1	+ .88	- -.20	- .30	+/- .55	+/- -.54	+/- .80	- .37	- .04	- .16
Is50			1	-.05	.25	.35	= -.24	+/- .80	= .18	= .17	= .10
Grain size				1	- -.36	- -.01	- .18	- -.08	- -.13	- -.05	- -.02
Grain shape					1	- .12	- -.60	+/- .49	- .36	- -.14	- .59
Dry density						1	- -.36	- .62	- -.25	- .50	- -.37
Porosity							1	- -.41	+ -.69	- .46	+/- -.60
E-modulus								1	- .02	- .33	- .05
RTI									1	+ -.88	+ .88
Vol.% abr.min.										1	+ -.83
UCS/BTS											1

Figure 29 Correlation matrix of rock property parameters. A +, +/- and - sign indicate respectively a good, moderate and bad relationship, not necessarily linear. Below the correlation coefficient is given

describe the behaviour of rock better than a set of dependent variables. From the correlation matrix different sets of independent parameters can be selected for investigation. A choice was made based upon a combination of criteria such as:
- the variation and measurement of a parameter should be easy.
- the parameter should be as basic as possible (i.e. not composed of other parameters).

The volume percentage of abrasive minerals, the grain size and shape and the material strength[24] are independent of one another, not composed of other para-

---

[24] Rock material strength can be expressed by the UCS value for the compressive strength and Is50 or BTS values for the tensile strength. These three parameters are related

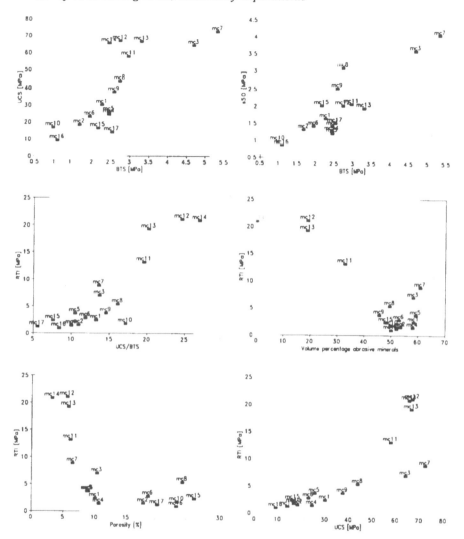

Figure 30 Scatter plots of some rock property parameters, which showed a moderate or good relationship

to another but it is not clear which parameter describes the strength during rock cutting best. Probably the compressive strength (UCS) describes the cutting behaviour of rock at the start of a cut (at a small feed; wear/cutting mode I) best, whereas the tensile strength (Is50 or BTS) describes the cutting behaviour of rock at larger feeds (wear/cutting mode III). At mode I the chisel only scrapes the rock producing an insignificant volume of rock cuttings and the rock (surface) is therefore mainly loaded in shear or compression. In mode III the chisels produces chips and according to some literature about rock cutting (briefly reviewed in chapter 3), the process of cutting rather than scraping relates best to the tensile strength of the rock. Both Is50 and BTS are indicators of the tensile strength. An advantage of the BTS over the Is50 is, that it also relates linearly to the dry density and the porosity. By this only BTS may be used instead of the Is50, porosity and dry density.

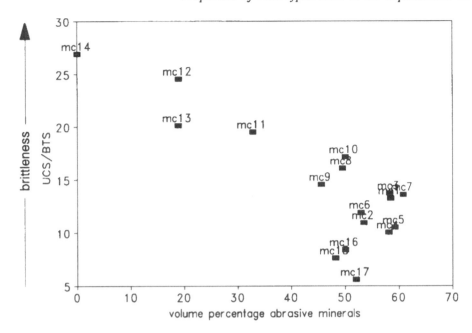

Figure 31 The ratio of UCS to BTS, which is considered an indicator of brittleness, versus the volume percentage of abrasive minerals in the mortars of Table 8

meters and were varied in the experiments one at a time. The other rock parameters are more or less dependent on one or more of the former rock parameters and are not so deeply considered.

9.2 ROCK SAMPLE PREPARATION

Rock samples were prepared to fit the experimental set-up. Rock cylinders, with a diameter of ± 140 mm were cored out of rock blocks. The cores were sawn into flat discs with a thickness of ± 80 mm. Grooves were sawn into the discs at both sides of the future cutting-path of the chisel to allow cracks to propagate to free surfaces through the rock at the sides of the chisel. The grooves simulate free surfaces, which will be created in rock cutting practice by other chisels, travelling at either side of the chisel.

CHAPTER 10

Test results on artificial rock (mortar)

10.1 REPEATABILITY AND THE EFFECT OF GROOVE SPACING

10.1.1 *Effect of spacing of the grooves, sawn into the rock disc*

Scraping experiments have been carried out with wedge steel type of chisels on three rock types: artificial lime-sandstones H, A and B (see chapter 9). The spacing between the grooves, sawn in the rock discs during sample preparation, was varied. An increase of the spacing between the grooves causes an increase of the width of cut and thus an increase of the volume of cut rock.

The tests were performed at low cutting velocities ($\pm$ 0.4 m/s) and at a relatively low feed ($\pm$ 0.14 mm per revolution).

Wear of the chisel increased with an increase of the width of cut. The shape of the worn chisels after testing was different for different widths of cut. A width of cut, equal to the width of the chisel (5 mm), caused the chisel to wear in such a way, that a concave wear-flat was formed (see Figure 32). A gradual increase of the width of cut caused a gradual change from a concave shape of the wear-flat into a convex shape of the wear-flat. In between the shape of the wear-flat is flat. The width of cut, at which the shape is flat, depends on the type of rock. It was about 10 mm for rock type H, and 7.5 mm for rock type B.

Hence the rock type clearly affects the shape of the chisel wear-flat. To minimize the shape alteration of the test chisel the groove spacing, which determines the width of cut, should be chosen in such a way that the wear-flat wears away resulting in a flat wear-flat shape.

10.1.2 *Repeatability of the experiments, with respect to the amount of wear*

Tests have been carried out to find the repeatability following BS 812:1975.

Identical scraping experiments have been performed with the same wear path length, cutting velocity (0.4 m/s), feed (0.14 mm/revolution) and width of cut (about 10 mm). Wedge steel type of chisels have been used.

In Table 10 the average, the standard deviation (St. dev.) and the repeatability

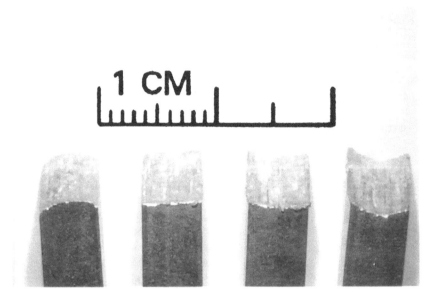

Figure 32 Photograph of the wear-flat shape from concave to convex (from right to left), caused by an increase of the width of cut (from about 5 mm to 15 mm)

(Repeat.) of the test settings and results are shown, based on five tests on artificial lime-sandstone type H.

Four tests on artificial lime-sandstone type A resulted in an average amount of chisel wear of 0.447 g and the repeatability of the weight measurement of chisel wear resulted in 0.029 g.

The repeatability should be low enough to be able to discriminate between experiments with different test settings. Thus, with respect to the results of the initial experiments on lime-sandstone A and H, the scraping test show a good repeatability.

10.2 SCRAPING TEST EXPERIMENTS ON MORTAR AT CONSTANT CUTTING VELOCITY

Experiments with the scraping test as described in chapter 7 have been performed on seventeen types of mortar of different compositions and with different properties (Table 8), using chisels made of two types of steel: dredge steel and wedge steel[25] (chapter 8). The experiments to be discussed in this paragraph were carried out at a constant cutting velocity of 0.4 m/s. The spacing between the sawn-in grooves, was 12 mm wide in all the experiments. The feed has been varied from 0.05 mm per

---

[25] If dredge steel chisels have been used, it will be mentioned explicitly in the text. If the type of steel is not mentioned, wedge steel has been used.

Table 10. Repeatability of test settings and test results. F_z is the average cutting force, F_y is the average normal force and F_x is the average lateral force on the chisel

	Velocity (m/s)	Cutting depth (mm)	Wear path (m)	F_z (N)	F_y (N)	F_x (N)	Chisel wear (g)
Average	0.419	0.147	46.54	90.7	118.7	5.6	0.312
St. dev.	0.0047 (1.1 %)	0.0018 (1.2 %)	0.032 (0.1 %)	6.5 (7.2 %)	8.1 (6.8 %)	1.2 (21 %)	0.012 (3.9 %)
Repeat.	0.013 (3.1 %)	0.0049 (3.3 %)	0.89 (0.2 %)	17.9 (20 %)	22.4 (19 %)	3.4 (61 %)	0.032 (10.3 %)

revolution (lowest feed possible in the range of the lathe) to 2.5 mm. The tests were performed as described in chapter 5. Parameters like specific energy, rate of wear and others (Table 3, chapter 7), which describe the rock cutting and wear processes were measured and calculated. These parameters were compared for various types of mortar to determine the effects of rock material properties like strength, porosity, mineral hardness, grainsize etc. on the rate and type of wear and on the cuttability. Finally some experiments were carried out at different cutting velocities.

Macro-photographs were made of chisel sections, which showed changes of the steel structure. At an increasing distance from the wear-flat, the Micro-Vickers Hardness of tested dredge steel chisels was measured. Changes in hardness and steel-structure are indications of heating occurring at the chisel wear-flat during testing. Specific colours of the chisel tip after a test also indicate temperatures reached at the wear-flat of the chisel during testing.

10.2.1 *Experiments on mortar composition 1*

10.2.1.1 *The rate of wear*
In Figure 33 the rate of wear of chisels made of dredge steel and chisels made of wedge steel is shown in experiments on mortar composition 1 (see Table 8) as a function of the feed. Each dot in the graph represents one test with fixed settings of all variables. The graph shows that for small feeds, the rate of wear increases rapidly with an increase of feed and reaches a maximum when the feed has a value of about 0.15 mm/revolution. When the feed is increased further the rate of wear decreases, first rapidly and thence, for feeds larger than 0.4 mm/revolution, slowly.

10.2.1.2 *The cutting modes*
The ratio of the radial displacement of the chisel into the rock (cutting depth) to the radial displacement of the support of the lathe (feed) may be a measure of different modes of cutting. This ratio is called the cutting mode ratio. Cutting mode ratios close to 0 are related to a scraping process or a low penetration of the chisel into the rock (cutting depth), relative to the feed. The difference between feed and cutting depth of the chisel is the rate of wear. In the extreme case the chisel wears away as fast as it is moved into the rock by the support of the lathe (cutting mode ratio is 0).

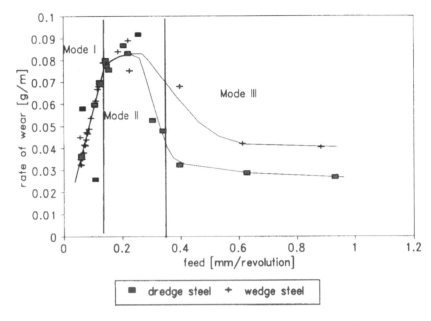

Figure 33 Rate of wear [g/m] versus feed [mm/revolution] for two steel types tested on mortar composition 1

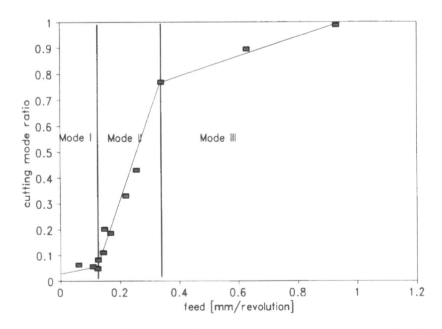

Figure 34 Cutting mode ratio versus feed [mm/revolution] for mortar composition 1

As a result the production of cut rock is zero and the amount of wear of a chisel per unit rock cut (specific wear) is high. Cutting mode ratios close to 1 indicate that largely a cutting process is taking place. Relatively much rock is cut by the chisel compared to the rate of wear. The specific wear is low.

In Figure 34, cutting mode ratios for tests at different feeds are shown for rock composition 1. The graph can be split into three areas, which are related to different cutting modes:
- Cutting mode I: rock is not breaking out in front of the chisel.
- Cutting mode II: rock partly breaks out in front of the chisel.
- Cutting mode III: rock breaks out in front of the chisel.

10.2.1.3 *The wear modes*
The cutting modes are linked to the wear modes described in chapter 3. The limits of the wear modes are drawn in Figure 33 and match the limits of the cutting modes.

In wear mode I the rock remains intact. Two-body wear prevails. High contact stresses develop between the asperities of the rock surface and the wear-flat of the tool, causing high temperatures. In the experiments high temperature phenomena have been observed at the wear-flat: high temperature colours, plastic deformation of the steel and burrs at the chisel edges appeared (Figure 35 and Figure 36). On the wear-flat grooves could be seen, which resulted from microcutting by rock surface asperities (Figure 38).

Wear mode II represents the transition of wear mode I to wear mode III. High temperatures appear in the first part of mode II. Higher contact stresses between the chisel and the rock causes higher temperatures than the temperatures reached in mode I. The prevalence of two-body wear close to wear mode I gradually changes to prevalence of three-body wear near wear mode III. Here, lower temperatures appeared, the amount of wear decreased and more crushed rock material was observed in the wear-flat of the tested chisels (Figure 39).

In wear mode III the rock relief is affected by the cutting action of the chisel. Three body wear occurs. Abrasive asperities and grains are crushed. No signs of high temperature phenomena, like tempering colours on the chisel or plastic deformation and burrs at the edges of the chisel, were observed in the experiments. Rock crushings were embedded into the wear-flat of the rock (Figure 37). These rock crushings could have protected the underlying steel against further wear.

10.2.1.4 *The specific wear and the specific energy*
The amount of wear of a chisel per unit volume of rock cut (specific wear SPW [kg/m^3]) is a measure of cutting efficiency with respect to wear of the chisels. In Figure 40 the relationship between SPW and feed is shown. Each dot in the graph represent one test run. The relationship holds for chisels of dredge steel and for chisels of wedge steel. The specific energy (SPE [MJ/m^3]) is a measure for the cutting efficiency for a specific rock cutting apparatus, in this case the scraping test set-up. The SPE is the amount of energy required to cut a unit volume of rock. In Figure 41 the SPE is shown for tests at different feeds. The SPE versus the feed shows the same trend as the SPW versus the feed and holds for wedge steel as well as for dredge steel. This implies that the cutting process and the wear process are related to each other.

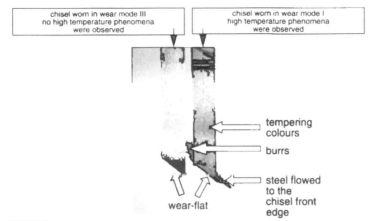

Figure 35 Two chisels, worn in different wear modes, showing different wear phenomena.

Figure 36 Wear-flat of a chisel, typically worn in wear mode I or the first part of wear mode II.

Figure 37 Wear-flat of a chisel, typically worn in wear mode III or in the second part of wear mode II.

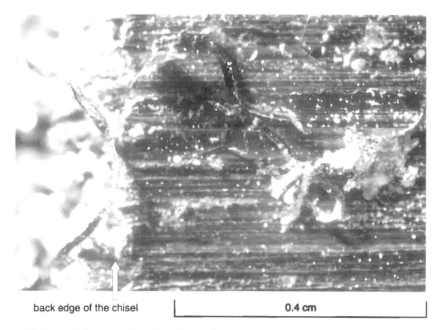

back edge of the chisel | 0.4 cm |

Figure 38 Part of the wear-flat of a chisel of dredge steel worn in wear mode I. Grooves and strings of gouged material are visible. Microcutting seems to be the material failure mechanism.

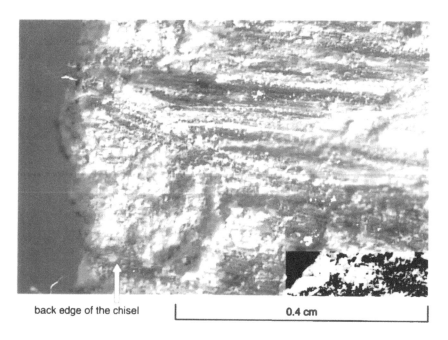

back edge of the chisel | 0.4 cm |

Figure 39 Part of the wear-flat of a chisel of dredge steel worn in wear mode III. Crushed rock material on and bedded into the chisel wear-flat. Gouges are most likely due to microcutting.

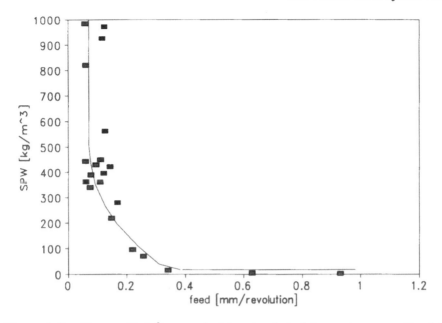

Figure 40 Specific wear [kg/m³] versus feed [mm/revolution] for mortar composition 1 (dredge steel chisels)

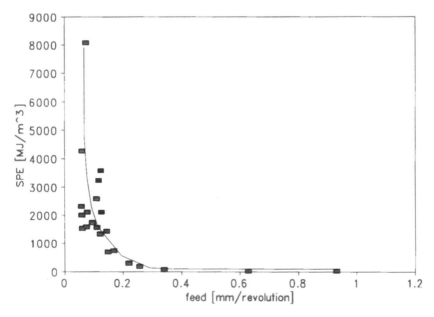

Figure 41 Specific energy [MJ/m³] versus feed [mm/revolution] for mortar composition 1 (dredge steel chisels)

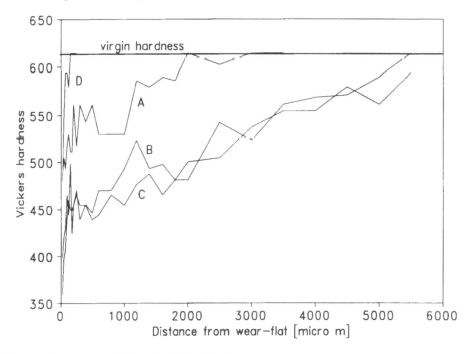

Figure 42 Hardness variation of chisels of dredge steel versus the distance from the wear-flat for chisels tested at different feeds

10.2.1.5 *Hardness determination and temperature estimation*

The wear behaviour of wedge steel and dredge steel is almost the same. The rate of wear in wear mode I and II is approximately the same for both steel types. This is probably caused by the presence of such high temperatures at the wear-flat of the chisel during testing, that the difference between the initial (virgin) hardness of wedge steel ($\pm$ 300 VH) and dredge steel ($\pm$ 600 VH) disappears during testing. In wear mode III however, the initial hardness is hardly affected (lower temperatures and contact stresses). Therefore, in this wear mode the rate of wear of wedge steel is higher than of dredge steel.

Micro vickers hardness measurements on sections of dredge steel chisels, tested at different feeds, have been carried out at increasing distances from the wear-flat. In Figure 42 can be seen that the hardness of the steel of the chisels, tested at a feed in wear mode I (chisel A, feed 0.11 mm/revolution) and in wear mode II (chisel B and C, feed 0.17 mm/revolution), is more influenced than the hardness of chisels tested at a feed in wear mode III (chisel D, feed 0.26 mm/revolution).

Photographs of etched chisel sections (Figure 43, Figure 44 and Figure 45) of the chisels C, A and D can be related to the hardness profiles in Figure 42. The initial structure of the steel, sorbite, is affected by high temperatures and contact stresses at the wear-flat during a period of 65 to 75 seconds. Differences in temperatures reached and contact stresses between various experiments cause differences in the amount of change of the initial steel structure.

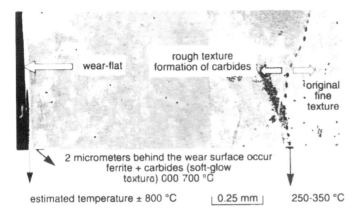

wear-flat

rough texture
formation of carbides

original
fine
texture

2 micrometers behind the wear surface occur
ferrite + carbides (soft-glow
texture) 000 700 °C

estimated temperature ± 800 °C | 0.25 mm | 250-350 °C

Figure 43 Polaroid photograph of the etched steel section of chisel
C, worn in wear mode II.

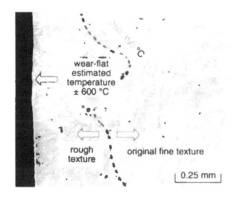

wear-flat
estimated
temperature
± 600 °C

250-350 °C

rough
texture   original fine texture

| 0.25 mm |

Figure 44 Polaroid photograph of the etched steel section
of chisel A, worn in wear mode I.

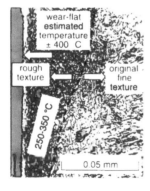

wear-flat
estimated
temperature
± 400 C

rough
texture

original
fine
texture

250-350 °C

| 0.05 mm |

Figure 45 Polaroid photograph of the etched steel section
of chisel D, worn in wear mode III.

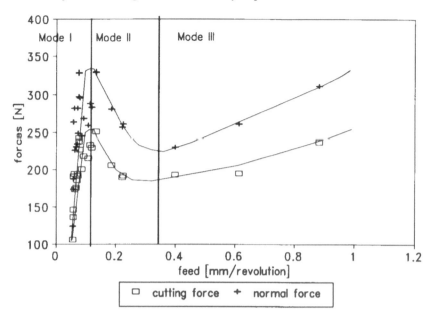

Figure 46 Cutting forces [N] versus feed [mm/revolution]. Results of scraping experiments using wedge steel and mortar composition 1

The transition of the initial fine texture of the sorbite into a rough texture appears in a temperature zone of 250 °C to 350 °C (Colijn pers. comm.). The photographs show the transition at about 2.5 mm from the wear-flat fore chisel C, at about 0.3 mm from the wear-flat for chisel A and at about 0.02 mm from the wear-flat for chisel D. A transition positioned at a larger distance from the wear-flat, may indicate that higher temperatures were present at the wear-flat, assuming that the chisels were subjected to equal heating periods. If temperature is the only parameter affecting the structure of the steel, chisel C has been subjected to the highest temperatures, followed by chisel A and finally chisel D. Estimated temperatures at the wear-flat are about 800 °C for chisel C, 600 °C for chisel A and 400 °C for chisel D.

However, as mentioned before, contact stresses also affect the steel structure. At the wear-flat of chisels worn in wear mode I and II a band of steel, approximately 2 micrometers in thickness, has been subjected to high contact stresses and smeared in the direction of movement of the chisel. As a result the steel structure is realigned and magnetized.

Due to the presence of both influencing factors it is not possible to deduct the precise temperatures and stresses, which prevailed at the contact points of chisel and rock in the experiments, from the changes in steel structure. Attempts to determine temperatures directly by using thermocouples failed because the effects of too many parameters, such as the area of contact between chisel and rock, were unknown or because the thermocouples installed near the wear-flat were damaged by the wear process. Readings of thermocouples installed further away from the wear-flat are only averaged values of temperature and are therefore not representative for the flash-temperatures at the contacts of rock and steel.

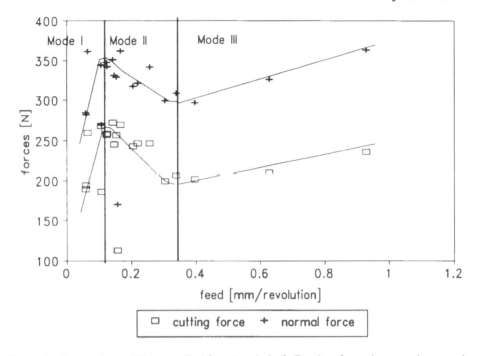

Figure 47 Cutting forces [N] versus feed [mm/revolution]. Results of scraping experiments using dredge steel and mortar composition 1

10.2.1.6 *The cutting forces*

In Figure 46 and Figure 47, the cutting forces are shown as a function of the feed for respectively wedge steel and dredge steel. The cutting forces include the cutting or tangential force and the normal or radial force. Each dot in the graphs represent the average cutting or normal force in a test run at fixed settings. The behaviour of the cutting forces with increasing feed is approximately the same for both steel types. At the end of mode II and in mode III, the normal force in the experiments with dredge steel (Figure 47) is higher than the normal force in case of experiments with wedge steel (Figure 46). The wear or cutting mode limits are drawn in the figures (vertical lines) and intersect maximum and respectively minimum values of the cutting force.

10.2.2 *Results of experiments on three mortars of different strength*

Next to the experiments on mortar composition 1, scraping test experiments have been carried out using samples of rock mortar compositions 2 and 3 (Table 8 and 11). The rock properties of the different mortars, like porosity, abrasive mineral content (quartz), grain size and grain shape are approximately the same except for the strength. The strength properties are different. Mortar composition 2 is the

Table 11. Composition and some properties of mc 1, mc 2 and mc 3 (see also chapter 9). UCS is the unconfined compressive strength. BTS is the Brazilian tensile strength. Volume percentage of abrasive minerals (in case of these mortars quartz) between the brackets are corrected for the porosity

mortar composi-tion mc	vol % abrasive minerals	grain size abrasive minerals (mm)	grain shape abrasive minerals	UCS (MPa)	BTS (MPa)	poro-sity (%)
mc 1	65 (58)	1.5	rounded	30	2.3	10
mc 2	65 (53)	1.5	rounded	18	1.7	18
mc 3	65 (58)	1.5	rounded	64	4.7	10

weakest, mortar composition 3 the strongest and mortar composition 1 has intermediate strength.

10.2.2.1 *The rate of wear*
Figure 48 shows that the rate of wear is determined by the rock material strength. A higher rock material strength causes a general increase of the rate of wear. Different rates of wear, related to the wear modes, can only be distinguished for mortar compositions 1 and 3. Almost all experiments on mortar composition 2 fall within wear mode III. Only experiments at small feed (<0.18 mm/revolution) fall within wear mode II, but the increase in the amount of wear is here too small to be observed. Still the wear mode transition at 0.18 mm per revolution could be established by observing the wear-phenomena on the wear-flats.

10.2.2.2 *The cutting and wear modes*
The type of cutting process and the type of wear is also affected by the rock material strength. An increase of the rock material strength clearly influences the position of the cutting mode limits with respect to the feed (Figure 49). Since cutting and wear mode limits are related, rock material strength also influences the position of the wear mode limits. A plot of those values of the feed at which a transition from one wear or cutting mode to another takes place versus the unconfined compressive strength (UCS) of the tested mortars is shown in Figure 50.

10.2.2.3 *The type of steel*
The difference in susceptibility to wear of the different steel types depends on the prevailing wear mode during the experiments. In wear mode I and partly in wear mode II, the rates of wear are equal for wedge and dredge steel. However, in wear mode III the rate of wear of dredge steel and wedge steel differs. In wear mode III, the rate of wear of dredge steel is approximately 75 % of the rate of wear of wedge steel. Since this holds for the experiments on mortar composition 1 as well as on mortar composition 3 (no experiments with dredge steel on mortar composition 2 were carried out), the difference in rate of wear of dredge and wedge steel is independent of mortar (rock) material strength.

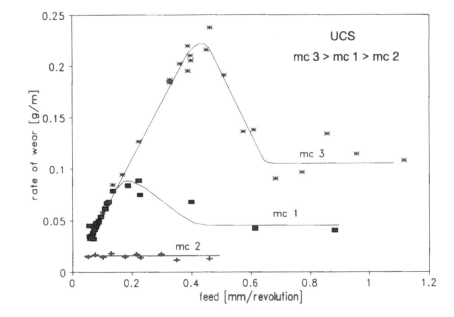

Figure 48 Rate of wear of chisels of wedge steel [g/m] versus feed [mm/revolution]. (mc 1, mc 2 and mc 3 are respectively mortar compositions 1 to 3)

10.2.2.4 *The specific wear and the specific energy*

The specific wear versus the feed is shown in Figure 52 for experiments on the three mortar types. The specific wear decreases with decreasing mortar material strength. Especially at low values of feed the difference of specific wear between the mortar types is large. Note the logarithmic scale, which was necessary to plot the mortar compositions in one figure. The initial increase of specific energy with increasing feed (only visible for mc 3) is caused by the increase of contact stresses and the subsequent rise of temperatures at the wear-flat of the chisel in wear mode I. A maximum of the SPW is reached when the rate of wear is high and the production of cut mortar material is still relatively low. Therefore the maxima should approximately be located at the limits of mode I to mode II, which are represented by the maxima of the curves drawn in the figure. The maximum of the curve of mc 2 is out of the feed range, which could be reached by the lathe. Maxima of the curves of mc 1 and mc 3 coincide reasonably well with the transition from wear mode I to II. The specific energy (SPE) versus the feed, shows the same trends as the SPW versus feed (Figure 53). Thus the cuttability of mortar (rock) decreases with increase of strength and/or decrease of the feed.

10.2.2.5 *The cutting forces*

The behaviour of cutting forces as a function of feed in experiments with the three mortar compositions is shown in Figure 51. Cutting forces include the cutting or tangential force (Fz) and the normal or radial force (Fy).

Fz is higher than Fy for all three mortars. Cutting forces are relatively high in experiments on mortar composition 3 (mc 3) and low in experiments on mortar

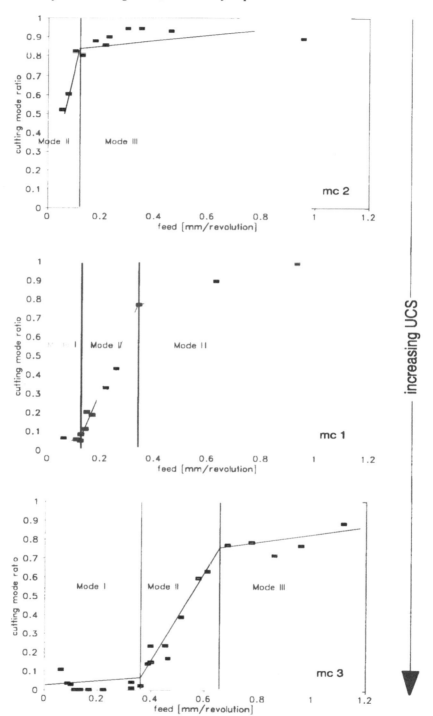

Figure 49 Cutting mode ratio versus feed [mm/revolution] for mortar compositions (mc) 1, 2 and 3

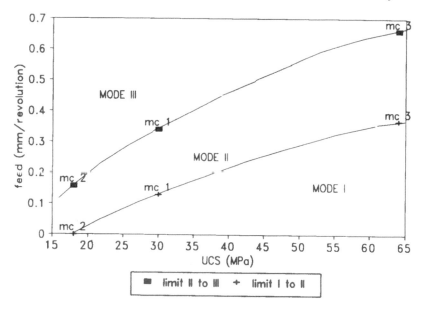

Figure 50 Three areas of different wear or cutting modes, related to the unconfined compressive strength of the mortar and the feed

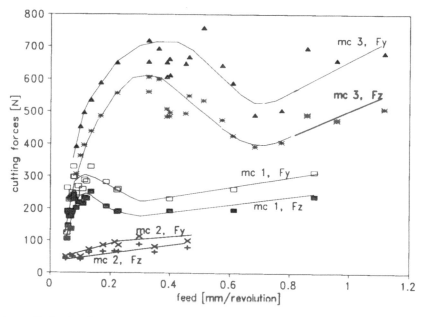

Figure 51 Cutting force (Fz) and normal force (Fy) for three mortar compositions (mc 1, mc 2, and mc 3) versus the feed [mm/revolution]

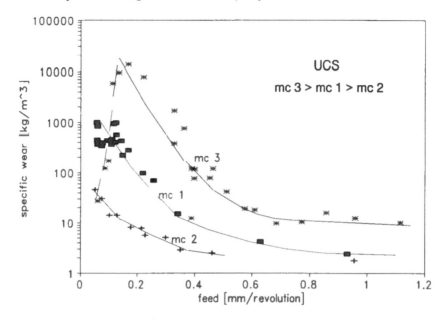

Figure 52 Specific wear [kg/m³] versus the feed [mm/revolution] for mortar compositions 1 to 3 (mc 1, mc 2 and mc 3)

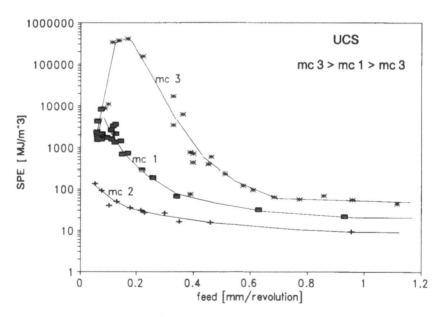

Figure 53 Specific energy [MJ/m³] versus the feed [mm/revolution] for mortar compositions 1,2 and 3

Table 12. Composition and some properties of mc 1, mc 4, mc 5 and mc 6 (see also chapter 9). UCS is the unconfined compressive strength. BTS is the Brazilian tensile strength. Volume percentage of abrasive minerals (in case of these mortars quartz) between the brackets are corrected for the porosity

mortar composition mc	vol % abrasive minerals	grain size abrasive minerals (mm)	grain shape abrasive minerals	UCS (MPa)	BTS (MPa)	porosity (%)
mc 1	65 (58)	1.5	rounded	30	2.3	10
mc 4	65 (58)	0.75	rounded	24	2.4	11
mc 5	65 (59)	0.38	rounded	26	2.4	9
mc 6	65 (53)	0.19	rounded	23	1.9	19

composition 2 (mc 2). The difference between the cutting forces is probably caused by the difference in strength of the tested mortars. Not only the height of the cutting forces is influenced by the strength, but also the positions of maxima and minima of cutting forces shift towards larger feeds when a stronger mortar is tested. This effect fits the effect of the shift of wear mode limits, when different mortar types are tested. Cutting forces in experiments on mortar composition 2 show a slightly different behaviour, since mainly cutting mode III occurs.

High temperature-phenomena at the wear-flat were therefore only observed during experiments on mortar compositions 1 and 3 (only in cutting or wear modes I and II).

10.2.3 *Results of experiments on four mortars of different grain size*

Experiments on mortar compositions 1,4,5 and 6 (see chapter 9, Table 8 and Table 12) have been carried out. All these mortars were composed of about 65 % rounded quartz grains (sand) and about 35 % portland B cement. The porosity of all these mortar types was about 10 %, except mc 6. The strength of the mortar types was about 25 MPa. Only mortar composition 1 was slightly stronger: 30 MPa. The mortar compositions differ in grain size (mc 1: 1.5 mm, mc 4: 0.75 mm, mc 5: 0.38 mm, mc 6: 0.19 mm).

10.2.3.1 *The rate of wear*
In Figure 54 it is shown that a larger grain size causes a higher rate of wear. The shapes of the curves of mc 1 and mc 4 differ from mc 5 and mc 6. The maxima in the rate of wear versus feed curve for mc 5 and mc 6 may be located at smaller values of feed than can be reached by the lathe or they may not exist at all. In case of mc 5 the transition from wear mode II to III is located at a feed of 0.12 mm/revolution based upon a change in wear phenomena at this feed.

10.2.3.2 *The cutting and wear modes*
The grain size clearly influences the position of the cutting mode limits (and the wear mode limits) with respect to the feed (Figure 55). An increase of the grain size of the

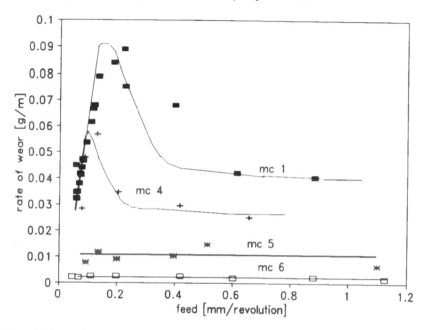

Figure 54 Rate of wear of chisels of wedge steel [g/m] versus feed [mm/revolution]. (mc 1, mc 4, mc 5 and mc 6 are respectively mortar compositions 1,4,5 and 6)

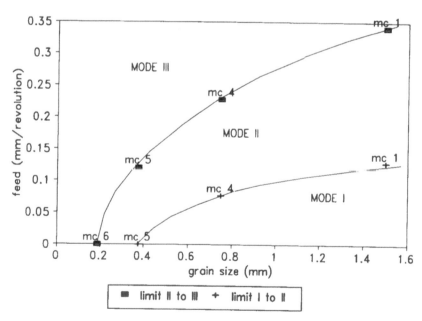

Figure 55 Three areas of different wear modes, related to the grain size of the mortar and the feed

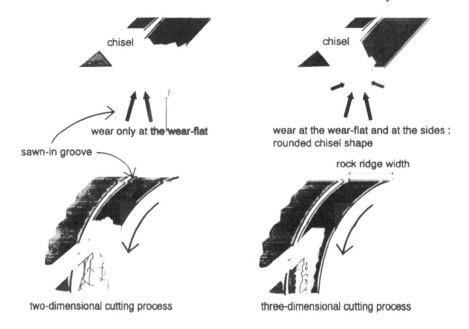

Figure 56 Illustration of a two- and three-dimensional cutting process. The transition occurs as a result of changing grain size of the rock

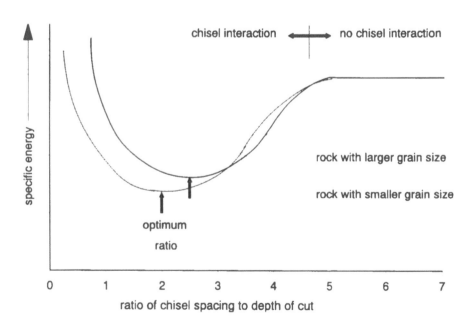

Figure 57 Specific Energy versus chisel spacing to depth ratio (Roxborough and Sen 1986). The optimum ratio (lowest specific energy) shifts to smaller spacing to depth ratio with decreasing grain size

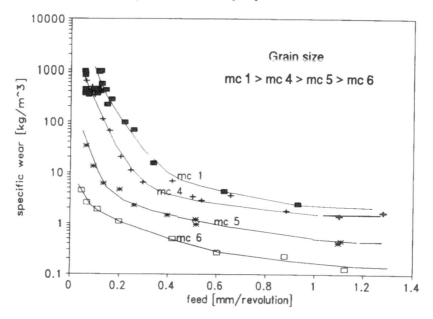

Figure 58 Specific wear [kg/m³] versus feed [mm/revolution] for mortar compositions 1,4,5 and 6

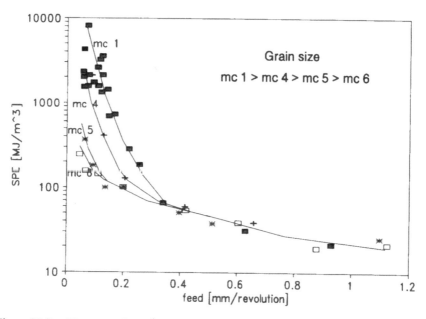

Figure 59 Specific energy [MJ/m³] versus feed [mm/revolution] for mortar compositions 1,4,5 and 6

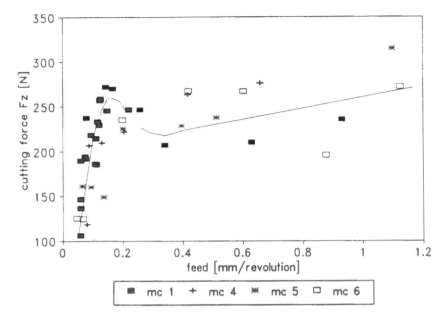

Figure 60 Cutting force (Fz,N) for mortar compositions mc 1, mc 4, mc 5, and mc 6 versus feed [mm/revolution]

mortar causes an increase of the feed at which transition from one wear mode to another takes place.

10.2.3.3 *The specific wear and the specific energy (the effect of chisel spacing)*

The rock ridge to be cut in the scraping test is defined by the grooves sawn into the disc during the sample preparation. The mortar fails by formation of rock chips by tension cracks which extend to the free surfaces, created by these grooves (Figure 56, two-dimensional cutting process). However, as the grain size of the rock decreases, the propagation of the cracks, which determine the size of the rock chips, changes: the direction of propagation and its extent decreases. As a result the width of the cut into the rock decreases with decreasing grain size. At a certain grain size the width of cut into the rock equals the set rock ridge width, defined by the sawn-in grooves. When rocks with even smaller grain sizes are used in the experiments, the cracks do not propagate to the free surfaces created by the sawn-in grooves (Figure 56, three-dimensional cutting process). Thus, the cutting process changes from a two-dimensional to a three-dimensional cutting process only by a decrease of grain size. In the three-dimensional case the rock-chisel contact does not only occur at the wear-flat, but also at the sides of the chisel, resulting in a rounded chisel shape. The same effect had also been observed as a result of changing the rock ridge width in the initial experiments (keeping all other variables constant) (Figure 32).

Roxborough and Sen (1986) show the relationship of the specific energy (SPE) with the ratio of pick (or chisel) spacing to cutting depth (Figure 57). The SPE is a measure of efficiency of cutting: lower values indicate a higher cutting efficiency. Chisel interaction plays an important role in this. Interaction of the chisels increases

Table 13. Composition and some properties of mc 3, mc 11 and mc 12 (see also chapter 9). UCS is the unconfined compressive strength. BTS is the Brazilian tensile strength. Volume percentage of abrasive minerals (in case of these mortars quartz) between the brackets are corrected for the porosity

mortar composition mc	vol % abrasive minerals	grain size abrasive minerals (mm)	grain shape abrasive minerals	UCS (MPa)	BTS (MPa)	porosity (%)
mc 3	65 (58)	1.5	rounded	64	4.7	10
mc 11	35 (33)	1.5	rounded	58	2.1	6
mc 12	20 (19)	1.5	rounded	67	2.0	6

the efficiency of cutting. At a certain ratio of chisel spacing to cutting depth, chisel interaction is optimal, i.e, the SPE has a minimum value. (Figure 57).

Scraping test experiments showed that crack propagation decreases when the grain size of the mortar decreases. Since chisel interaction exists by virtue of crack propagation, the range of chisel spacing to cutting depth ratios at which chisel interaction occurs shifts to smaller ratios as the grain size decreases (Figure 57). To achieve an efficient cutting process in practice while cutting in rocks of smaller grain size, chisels should be positioned at closer spacings on the cutter head. As a result more chisels are needed to cut a unit volume of rock with smaller grain size efficiently.

The relationship of SPW and feed for the mortar compositions 1,4,5 and 6 is shown in Figure 58. The SPW decreases with decreasing grain size and with increasing feed. In Figure 60 the SPE versus feed curve is shown.

The SPE, like the SPW, decreases with an increase of the feed. Mortars with larger grain sizes show a higher SPE, but only in experiments at small feeds (in the range of wear mode I and II). At larger ranges of the feed the SPE in experiments on different mortars was approximately equal. Grain size therefore seems to have only a limited effect on the cuttability of rock by a single chisel at larger feeds.

10.2.3.4 *The cutting forces*
The magnitude of the cutting forces in the experiments was not much affected by variation of the grain size. In Figure 60 the cutting force Fz is shown as a function of the feed for mc1, mc4, mc5 and mc6. The same holds for the normal force Fy.

10.2.4 *Results of experiments on three mortars of different volume percentage of abrasive grains (quartz)*

If the ratio of compressive to tensile strength is considered to be a measure of the brittleness of rock (Gehring 1987), the mortar became more brittle when the volume percentage of cement was increased (see Table 8 and Figure 31). As a result mortar samples containing higher volume percentages of cement became more sensitive to damage, like spalling of the edges of the mortar cores during sample preparation (boring and sawing). Scraping tests on mc 13 and mc 14 failed due to failure of the

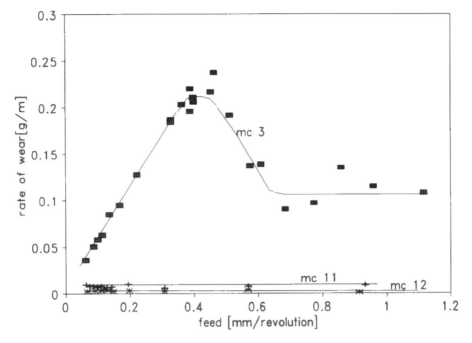

Figure 61 Rate of wear of chisels (wedge steel) [g/m] versus feed [mm/revolution] for mortar compositions mc 3, mc 11, mc 12

sample before the end of a test run. The increased brittleness may have its effect on the cuttability of the rock: the geometry of the mortar chips and the orientation of the planes along which the chips develop during cutting may alter with an increase of brittleness.

The composition and the properties of the used mortar compositions (mc 3, mc 11 and mc 12) are described in Tables 8 and 13. The three mortar compositions have approximately the same unconfined compressive strength. Mc 3 has a higher Brazilian tensile strength. The major differences between the mortar compositions are the volume percentages of matrix (cement) and quartz grains (sand); grain size 1.5 mm). Mc 3 contains 65 vol.% of quartz grains, mc 11 contains 35 vol.% of quartz grains and mc 12 contains 20 vol.% of quartz grains.

10.2.4.1 *The rate of wear*
The rate of wear decreases when the volume percentage of matrix (cement) increases (Figure 61).

10.2.4.2 *The cutting and wear modes*
Not only the rate of wear but also the type of wear (wear mode) is influenced by the volume percentages of quartz grains and matrix. Only mc 3 shows three ranges of feed where different wear modes occur. Mc 11 shows the transition of wear mode II to III at a low value of feed. Mc 12 only shows wear mode III. In Figure 62 curves representing values of feed at which a transition of wear modes takes place

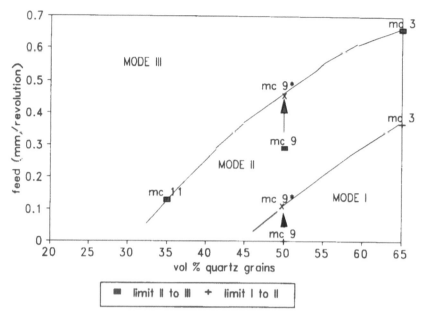

Figure 62 Three areas of different wear and cutting modes, related to the volume % of quartz grains of the mortar and the feed

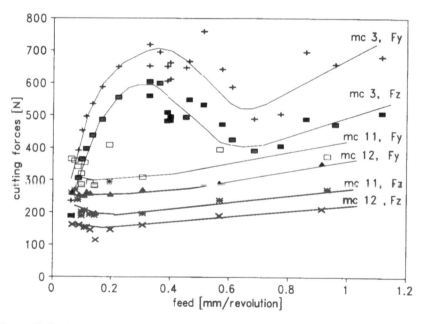

Figure 63 Cutting force (Fz,[N]) and normal force (Fy,[N]) for three mortar compositions (mc 3, mc 11 and mc 12) versus feed [mm/revolution]

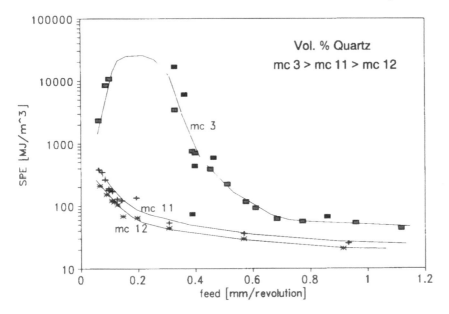

Figure 64 Specific energy (SPE) [MJ/m³] versus feed [mm/revolution] for mortar compositions which differ in quartz content (vol.%)

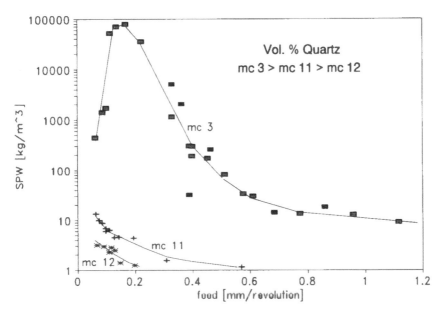

Figure 65 Specific wear (SPW) [kg/m³] versus feed [mm/revolution] for mortar compositions which differ in quartz content (vol.%)

Table 14. Composition and some properties of mc 2 and mc 15 (see also chapter 9). UCS is the unconfined compressive strength. BTS is the Brazilian tensile strength. Volume percentage of abrasive minerals (in case of these mortars quartz) between the brackets are corrected for the porosity

mortar composition mc	vol % abrasive minerals	grain size abrasive minerals (mm)	grain shape abrasive minerals	UCS (MPa)	BTS (MPa)	porosity (%)
mc 2	65 (53)	1.5	rounded	18	1.7	18
mc 15	65 (48)	1.5	angular	16	2.2	23

are shown. To draw the curves an extra mortar composition, mc 9, is used. Mc 9 contains 50 vol.% of quartz grains of 1.5 mm diameter. However the unconfined compressive strength (UCS) is only about 37 MPa. A correction for the low UCS of mc 9 has been applied to the feed at which a transition of the wear modes occurs. Mc 9 corrected is represented by mc 9* in Figure 62. This correction is deduced fromfigure 34.

10.2.4.3 *The specific wear and the specific energy*
The specific wear and the specific energy, SPW and SPE, as a function of the feed are respectively shown in Figure 65 and Figure 64. The graphs show a decrease of the SPW and the SPE with an increase of feed or/and decrease of volume % of quartz.

10.2.4.4 *The cutting forces*
In Figure 63 the cutting forces versus feed are shown for three mortar compositions. Although the three mortar compositions have approximately the same unconfined compressive strength (about 60 MPa), the forces, necessary to cut the rock, differ in magnitude. An increase of the volume percentage of matrix causes a decrease of the cutting forces. Or, as explained before, an increase of the brittleness of the mortar causes a decrease of the cutting forces. Cutting forces are a measure for the cuttability of rock. Therefore, the cuttability of mortar (or rock) increases with increasing brittleness or matrix content and the unconfined compressive strength alone is not a good indicator of the cuttability of rock. However, the Brazilian tensile strength correlates quite well with the cutting forces in mode III and may therefore be a better indicator of the cuttability of rock (or mortar) than the unconfined compressive strength is (Deketh 1993).

10.2.5 *Results of experiments on two mortars of different shape of the quartz grains*

Experiments on mortar compositions mc 2 and mc 15 (chapter 9, Table 8 and Table 14) are compared with each other in order to study the possible effect of the shape of the quartz grains on the rate of wear and the cutting forces for different feeds. The unconfined compressive strength of mc 2 and mc 15 is about 17 MPa. However, the

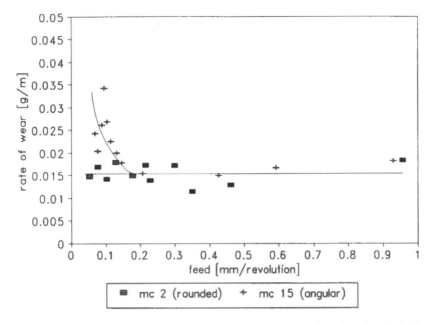

Figure 66 Rate of wear of chisels of wedge steel [g/m] versus feed [mm/revolution] for mortar compositions mc 2 and mc 15

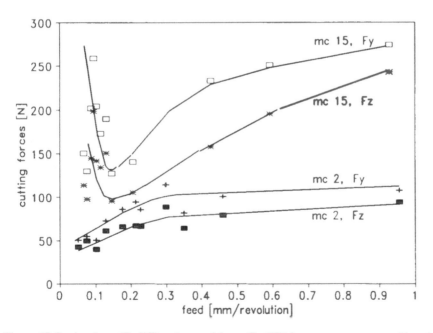

Figure 67 Cutting force (Fz,[N]) and normal force (Fy,[N]) for two mortar compositions (mc 2 and mc 15) versus feed [mm/revolution]

tensile strength is higher in case of mc 15 (BTS of mc 15 is 2.2 MPa and the BTS of mc 2 is 1.7 MPa). Also the porosity of mc 15 is higher.

10.2.5.1 *The rate of wear*
In Figure 66 the rate of wear versus the feed is shown for both mortar compositions. At feeds larger than about 0.18 mm/revolution the rate of wear for both mortar compositions is approximately equal. However, at values of feed smaller than 0.18 mm/revolution, the rate of wear in the experiments with mc 15 was higher than in the experiments with mc 2.

10.2.5.2 *The cutting and wear modes*
Not only the rate of wear but also the type of wear differs at small feeds. After testing mc 15 high temperature colours and burs were sometimes observed on the worn chisels, pointing to wear mode I or II. Chisels used for experiments on mc 2 did not show these phenomena. In the range of feeds tested, in case of mc 2 only wear mode III occurred. In case of mc 15, wear mode II prevailed up to a feed of ± 0.18 mm/revolution. At larger feeds wear mode III prevailed. Since in wear mode III the initial shape of the grains at the rock chisel interface is altered by the crushing, the effect of the shape on the type and rate of wear is considered to be less or absent at feeds where wear mode II occurred. This may explain the fact that the rate of wear in the experiments at feeds larger than 0.18 mm/revolution is approximately equal for both mortars.

Wear modes occurred at a different range of feeds for both mortar compositions. This is probably due to differences in tensile strength and/or in grain shape.

10.2.5.3 *The cutting forces*
In Figure 67 the cutting forces versus the feed for mc 2 and mc 15 are shown. The cutting forces in experiments with mc 15 were higher than those in experiments with mc 2, despite the fact that the unconfined compressive strength of both samples is approximately the same. Apparently the unconfined compressive strength is not a good indicator for the cutting forces. However, the Brazilian tensile strength may be related to the cutting forces. This corresponds with the conclusion drawn from experiments on mortar compositions containing different volume percentages quartz grains (chapter 10.2.4).

10.2.6 *Results of experiments on mortars of different mineral content*

Mortar compositions mc 16 and mc 17 contained glass or feldspar grains instead of quartz. The glass consisted of fused silica (99 vol.% of SiO_2) and the feldspar was a potash-feldspar retrieved from an alkali-feldspar granite. Both have Mohs hardness 6 (Mohs hardness of quartz is 7). The equivalent Vickers hardness is about 750 VH, which is only slightly higher than the virgin hardness of dredge steel.

Due to the possible occurrence of cleavage in the feldspar, the Vickers hardness may even be lower. The wear processes observed in the scraping test experiments on mc 16 and mc 17 were the same as for comparable quartz bearing mortars. Mc 16 and mc 17 have been compared with quartz bearing mortars of about the same

Table 15. Composition and some properties of mc 2, mc 10, mc 15, mc 16 and mc 17 (see also chapter 9, Table 8). UCS is the unconfined compressive strength. BTS is the Brazilian tensile strength. Volume percentage of abrasive minerals (in case of these mortars quartz) between the brackets are corrected for the porosity

mortar composition mc	vol % abrasive minerals	abrasive mineral	grain size abrasive minerals (mm)	grain shape abrasive minerals	UCS (MPa)	BTS (MPa)	porosity (%)
mc 2	65 (53)	quartz	1.5	rounded	18	1.7	18
mc 10	65 (50)	quartz	1.5	rounded	17	1.0	23
mc 15	65 (48)	quartz	1.5	angular	16	2.2	26
mc 16	65 (50)	glass	1.5	angular	9	1.1	23
mc 17	65 (52)	feldspar	1.5	angular	14	2.5	20

properties (mc 2, mc 10 and mc 15, Table 15). The Brazilian tensile strength of these mortars is not constant. Only mc 16 has a very low unconfined compressive strength compared to the other mortars. Mc 2 and mc 10 have been used to verify the possible influence of the Brazilian tensile strength to the specific wear in the strength range of these mortars. Mc 15 has been used because it contains angular grains and could therefore be compared to mc 16 and mc 17 as far as this aspect is concerned.

10.2.6.1 *Feldspar versus quartz*
In Figure 68 the SPW and in Figure 69 the SPE of wedge steel versus the feed is shown for mc 17 (with angular feldspar), mc 2 and mc 10 (with rounded quartz grains) and mc 15 (with angular quartz grains).

The difference in Brazilian tensile strength between mc 2 and mc 10 does not seem to manifest itself in a large difference in the specific wear in mode III. Experiments on both mortars yielded approximately the same specific wear at the same feeds. Thus experiments on mc 15 can be compared with experiments on mc 17 (at least at large feed values).

Although due its lower hardness, feldspar is considered to be less abrasive than quartz, the specific wear measured in the experiments was at least as high as observed in experiments on the mortars containing quartz sand. The test data found for dredge steel showed approximately the same results and therefore the same may be concluded for dredge steel experiments.

10.2.6.2 *Glass versus feldspar*
Mc 16, contains glass, which has a Mohs hardness of 6 (the same as the feldspar of mc 17). Feldspar is considered more susceptible to crushing, because of the possibility of cleavage and therefore it is considered to have a lower abrasive capacity than glass. Mortar compositions mc 16 and mc 17 cannot be compared directly because they differ too much in strength. The unconfined compressive strength of mc 16 is approximately half the strength of mc 17. In Figure 70 the specific wear of mc 16 and mc 17 versus the feed is shown. Despite the lower unconfined compressive strength of mc 16, the specific wear at small feeds (wear

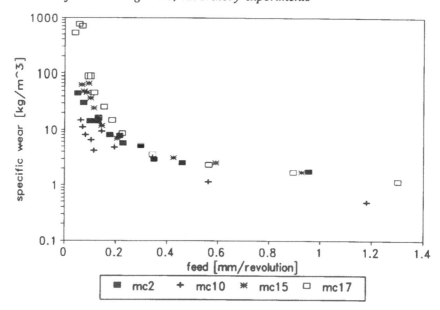

Figure 68 Specific wear SPW [kg/m³] versus feed [mm/revolution] for four mortar compositions. Mc 17 contains feldspar and the others bear quartz

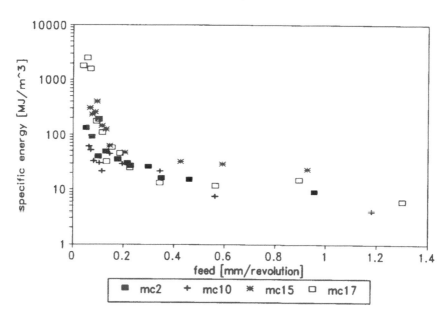

Figure 69 Specific energy SPW [MJ/m³] versus feed [mm/revolution] for four mortar compositions. Mc 17 contains feldspar and the others bear quartz

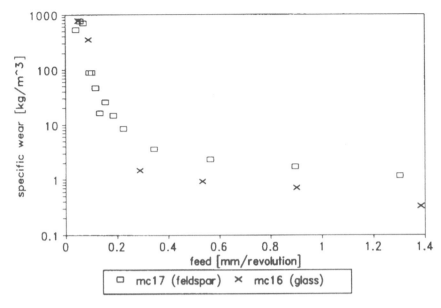

Figure 70 Specific wear [gk/m³] of mc 16 and mc 17 versus the feed [mm/revolution]. Mc 16 contains glass and mc 17 feldspar

modes I and II, feed smaller than 0.2 mm/revolution) is almost equal to the specific wear measured in the experiments on mc 17. This may be due to the different mineral content. The lower unconfined compressive strength of mc 16 is probably compensated by the lower friability (higher abrasive capacity) of the glass.

10.3 SPECIFIC WEAR EQUATIONS BASED UPON SCRAPING TEST EXPERIMENTS ON MORTAR

Based on the correlation analysis in chapter 9, four mutually independent rock material properties (These were: the volume percentage of the abrasive grains, the grain size and shape of the abrasive grains and the material strength) were considered important to the wear process and were therefore varied one at a time (section 10.1). The effects of these rock material properties on the rate and type of wear and on the specific wear and energy were visualized in and described by the graphs in this section.

By performing a regression analysis on the experimental results the combined effect of these rock material properties can be investigated. The Brazilian tensile strength (BTS) has been chosen as the main strength parameter in the regression analysis of the specific wear in wear mode III, since it incorporates also the influence of the bulk density and porosity and relates better to the cutting force than the unconfined compressive strength (UCS) does in wear mode III (Deketh 1993). Since the point load index (Is50), the ratio of UCS to BTS and the rock toughness index (RTI) relate well to one or more of the four rock material parameters under investigation (chapter

9), no further attention will be paid to them. However, the porosity, the bulk density, the UCS and the elasticity modulus (E-modulus) relate only to a certain extent with the four rock material parameters under investigation and should therefore not be completely ignored. Only minor attention will be paid to them in this research, but they should be kept in mind when analyzing the influences of the four investigated rock material parameters on wear or rock cutting processes. Apart from the parameters discussed also the type of abrasive mineral may play a role in the wear processes. However, it is less clear which parameters determine the behaviour of minerals. Often mineral properties are affected by weathering and alteration. As an example of the effect of the type of mineral on the wear processes, feldspar was used in some scraping test experiments described in section 10.2.6.

It became clear, that regression[26] could best be carried out in two separate ranges of feed. The first range being the range of feed at which wear modes I and II prevail and the second range being the range of feed at which wear mode III prevails. The wear processes in the different wear modes differ from each other and relate therefore also differently to the rock material parameters. Equation (4) has been deduced from test data in wear modes I and II of mortar compositions 1 to 15. It describes the specific wear of wedge steel chisels versus the feed for wear modes I and II. The shape of the grains turned out not to be significant.

$$log SPW_{I\,and\,II} = -6.554 +0.054 UCS +0.108 vol.\% \; ab. \; min. +1.489\sqrt{\Phi} -5.5 feed \; [\frac{kg}{m^3}] \qquad (4)$$

Equation (5) describes the specific wear of wedge steel chisels on mortar compositions 1 to 15 for wear mode III.

$$SPW_{III} = -0.026 BTS +0.016 vol.\% ab.min. +(0.0078 * BTS * \Phi * vol.\% ab.min.)^2 \; [\frac{kg}{m^3}] \qquad (5)$$

The UCS and the BTS are expressed in MPa and the grain size (Φ) in mm. Instead of the volume percentage of quartz, the volume percentage of abrasive minerals (vol.%.ab.min.) should be used in the Equations to stress the possible abrasive nature of hard minerals[27] other than quartz. The volume percentages of the solid constituents together with the porosity make 100%. Scraping test experiments on a feldspar containing mortar showed the specific wear of wedge steel chisels to be the same as found otherwise for an identical quartz containing mortar (section 10.2.5). Abrasive minerals softer than quartz should therefore be taken into account when the SPW is calculated with the Equations (4) and (5).

When Equation (4) is added to Equation (5) this results in the specific wear for the

---

[26] Multi variable linear regression has been carried out using the computer programme ESM (Eenvoudige Statistische Manipulator) produced by van Soest, Faculty of Mathematics, Delft University of Technology. The significance criterion for the used parameters was 5%.

[27] A mineral is considered abrasive, if its hardness is harder than the material of the chisel.

complete feed range tested (from 0.05 to about 1.4 mm/revolution). Addition of the Equations is allowed since SPW values found with Equation (4) can be neglected at feed ranges in wear mode III and SPW values found with equation 5 can be neglected at feed ranges in wear modes I or II. The result of the addition of Equations (4) and (5) is Equation (6).

$$SPW = A*10^{-5.5 feed} + B \qquad [\frac{kg}{m^3}] \qquad (6)$$

in which

$A = 10^{-6.554 +0.054 UCS +0.108 vol.\% ab.min. +1.489\sqrt{\phi}}$

and

$B = -0.026 BTS +0.016 vol.\% ab.min. +(0.0078*BTS*\Phi*vol.\% ab.min.)^2$

In Figure 71 the graphs produced from this Equation show a good fit with the original test data (the dots in graphs b, c and d). At very small values of feed (smaller than 0.05 mm/revolution) Equation (6) yields higher SPW values than the experimental data. SPW values should therefore not be calculated in this range of feeds.

In graph a all measured SPW values are plotted versus the calculated. Only the predicted SPW values for mortar composition mc 7^{28} are too high and so Equation (6) gives therefore a somewhat conservative value. Disregarding the data on mc 7, the experimental SPW values versus the predicted SPW values correlate with a correlation coefficient r^2 of 0.82.

10.4 SCRAPING TEST EXPERIMENTS ON MORTAR WITH VARYING CUTTING VELOCITY

Schimazek (1970) and others found that above a certain "cutting" velocity named the critical velocity (V_{cr}), the rate of wear of the tested "chisels" increased rapidly. The cause of this increase of the rate of wear is probably the rise in temperature at the wear-flat with increase of the cutting velocity. At the V_{cr} the temperature is so high (T_{cr}), that the (hardened) steel structure collapses causing a decrease of the wear resistance of the chisel. The V_{cr} is different for different steel types and is also influenced by the stress regime. Besides this, the V_{cr} is different for different rock types. Schimazek (1970) found a relationship between the V_{cr} and the F-value (chapter 6, Equation (1)), which incorporates grain size, equivalent volume percentage of quartz and the tensile strength (Equation (7)). This Equation was based upon pin-on-disc experiments on Carboniferous rock types.

In the present work experiments on most of the mortar types listed in Table 8 were carried out at different cutting velocities (0.3-3 m/s) to observe how some rock material properties may influence the V_{cr}. Most experiments were performed with chisels made of dredge steel. Since the type of steel is supposed to influence V_{cr}, some experiments were also carried out with wedge steel. The feed was kept constant

---

[28] The reason for this discrepancy is not clear.

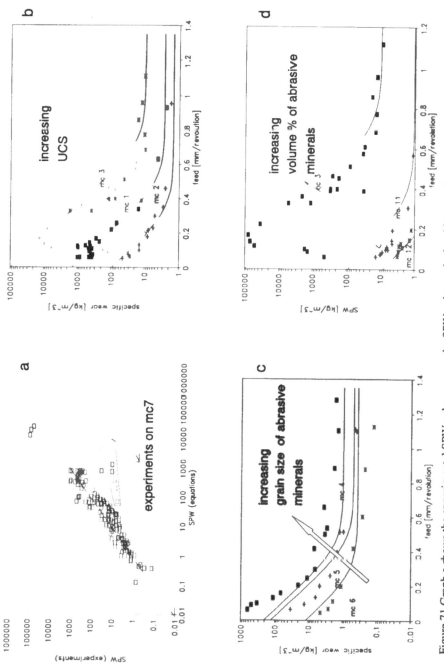

Figure 71 Graph a shows the experimental SPW values versus the SPW values calculated by equation 6. Graphs b, c and d show the fit with different mortar types

$$V_{cr} = k \ e^{-F}$$

k = *Constant which incorporates the chisel geometry*
  and the critical temperature of the chisel steel
e = *Base of the natural logaritm*
F = *Schimazek's wear value* [N/m]

(7)

at about 0.65 mm/revolution in all experiments. The dominant wear mode at this value of feed was wear mode III, in which contact temperatures are low at a cutting velocity of about 0.4 m/s. A low temperature at a low cutting velocity is a prerequisite for successful testing if the velocity at which the wear rate increases (V_{cr}), where the temperatures become so high that the steel structure of the chisel collapses, is to be found at higher velocities.

10.4.1 *Experiments on mortar composition 8*

Scraping test experiments on mc 8 (see Table 8) have been performed with dredge steel and wedge steel chisels.

10.4.1.1 *The specific wear*
The specific wear of dredge steel and wedge steel increased at high cutting velocities in tests on mc 8 (Figure 72). Mc 8 was used as an illustration of the observed effects because for this mortar the V_{cr} has a value approximately in the middle of the range of cutting velocities set in the experiments. The increase of the SPW is most likely caused by changes of the steel structures due to heat development at the wear-flats of the chisels. High temperature phenomena, burrs and tempering colours occurred at the wear-flats of the tested chisels.

While both steel types show an increase of the specific wear, dredge steel shows a more abrupt and larger increase than wedge steel does at cutting velocities higher than ± 1.5 m/s. (Higher forces are not related to higher rates of wear in experiments with low temperatures (low velocity and high feed) in wear mode III (section 10.2.1.2, Figure 33).

10.4.1.2 *The cutting forces*
Recorded forces in the experiments on mortar composition 8 are shown in Figure 73 for wedge and dredge steel.For cutting velocities lower than V_{cr} (1.5 m/s in case of mc 8), the forces remain approximately constant. At velocities higher than V_{cr} the forces increase.

The rate of increase of the cutting forces is smaller for wedge than for dredge steel, probably related to the difference in cutting mode ratio between tests with dredge steel and wedge steel chisels.

10.4.1.3 *The cutting mode ratio*
As has been explained in section 10.2.1.2 cutting mode ratios close to 1 indicate that a real cutting process is taking place, whereas cutting mode ratios close to zero

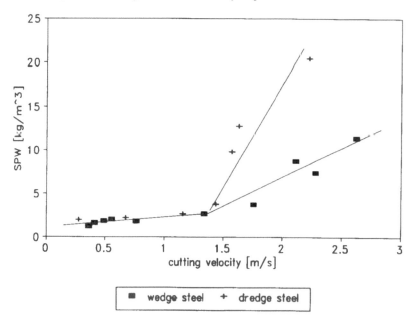

Figure 72 The specific wear [kg/m³] versus cutting velocity [m/s] for dredge steel and wedge steel on mortar composition 8; the feed is ± 0.65 mm/revolution·

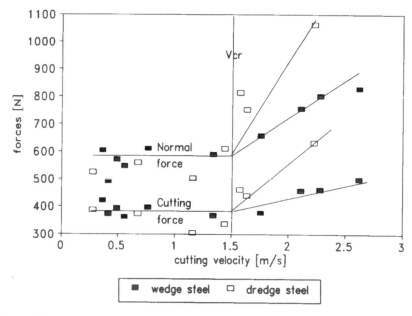

Figure 73 The cutting forces [N] versus the cutting velocity [m/s] for wedge steel and dredge steel on mortar composition 8; feed ± 0.65 mm/revolution

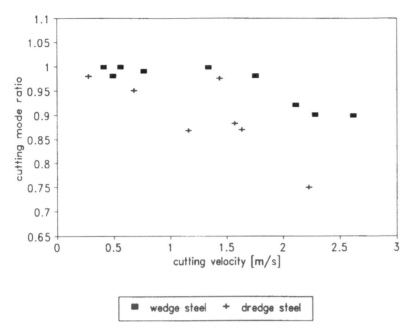

Figure 74 The cutting mode ratio versus the cutting velocity [m/s] for mortar composition 8; feed ± 0.65 mm/revolution

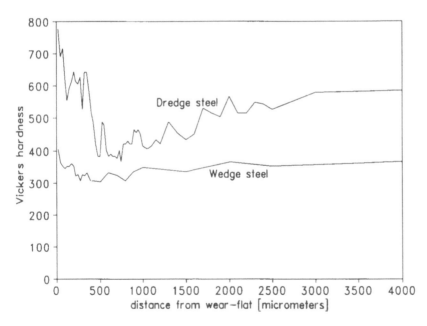

Figure 75 Vickers hardness measurements on sections of a wedge steel and a dredge steel chisel, which had been tested at a feed of 0.7 mm/revolution and at a cutting velocity of 2.1 m/s on mc8

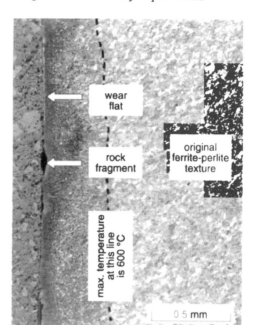

Figure 76 Polaroid photograph of the etched steel section of a wedge steel chisel tested on mc 8 at a feed of ± 0.7 mm/revolution and at a cutting velocity of ± 2.1 m/s.

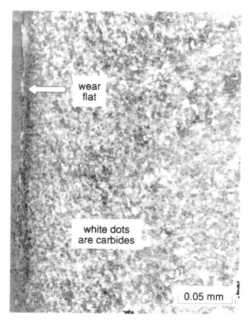

Figure 77 Polaroid photograph of the same chisel as in photograph 12, but the magnification is 10 times higher.

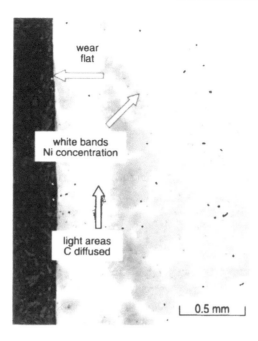

Figure 78 Polaroid photograph of the etched steel section of a dredge steel chisel tested on mc 8 at a feed of ± 0.7 mm/revolution and at a cutting velocity of ± 2.1 m/s

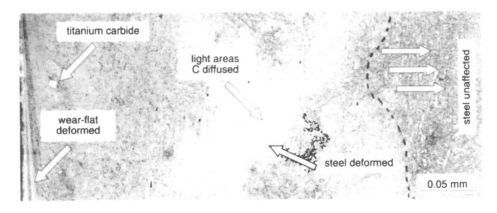

Figure 79 Polaroid photograph of the same chisel as in photograph 14, but the magnification is 10 times higher

indicate that merely a scraping process is taking place. The cutting efficiency will decrease for lower cutting mode ratios. At cutting velocities higher than V_{cr} the cutting mode ratio decreases for wedge steel as well as for dredge steel (Figure 74). The decrease of this ratio is larger for dredge than for wedge steel. At higher cutting velocities the cutting process changes to a scraping process. This is closely related to the increase in cutting forces and the increase of the rate of wear.

10.4.1.4 *The specific energy*
The specific energy increases at cutting velocities higher than V_{cr} because:
  1. the cutting forces increase
  2. the cutting efficiency (cutting mode ratio) decreases

10.4.1.5 *The role of texture of dredge steel and wedge steel*
Figure 72 shows that wedge steel appeared to be less susceptible to wear at cutting velocities higher than V_{cr} than dredge steel. Analysis of the texture of both steel types after testing may give a clue to their different response during testing. Micro photographs were made of etched sections of a dredge steel and a wedge steel chisel, which had been tested at a cutting velocity of $\pm$ 2.1 m/s (higher than the V_{cr}) on mc 8 at a feed of about 0.65 mm/revolution (Colijn pers. com.).

Wedge steel originally has a low temperature moulded ferrite-perlite texture (Figure 76, magnification 50 x). The white bands, which lie around the darker banded perlite, indicate the ferrite. This texture is stable at temperatures lower than 600 °C. Near to the wear-flat, this texture is changed by the action of high temperatures and deformation at the wear-flat. The texture becomes finer towards the wear-flat. Rock fragments with a size of about 50 to 100 μm can be seen bedded into the steel at the wear-flat. Figure 77 is a magnification of the area near the wear-flat in Figure 76. The cementite bands in the perlite are transformed and grown into the spherical carbides. Some remains of the ferrite bands can still be seen.

Dredge steel originally has a sorbite texture (Figure 78, magnification 50 x). The alloying element Ni is concentrated in the white dendritic pattern. Towards the wear-flat the original fine texture coarsens due to high temperatures and deformation of the steel at the wear-flat. In the pale zone (light areas), which is approx. 150 μm wide, the carbon (C) has disappeared from the steel near the wear-flat. In Figure 79, magnification 500 x, some details can be seen, like the plastic deformation of the steel near the wear-flat and the presence of titanium carbides, which remain stable even close to the wear-flat.

The differences in the texture of the two steel types which may be the cause that dredge steel has a lower wear resistance than of wedge steel, are listed below.

1. The thickness of the zone which has been influenced by high temperature and/or deformation was much larger in case of dredge steel than for wedge steel. This is probably due to the lower thermal conductivity of dredge steel caused by the presence of alloying elements. A low thermal conductivity of the steel can cause high temperatures at the wear-flat, since the rate of cooling is low. High temperatures, in their turn, can soften the steel and thus increase its susceptibility to wear.

2. Near the wear-flat, dredge steel has been decarbonized, causing a higher susceptibility to wear. Conversely, in wedge steel the cementite bands in the perlite have been transformed into carbides, which increase the wear resistance.

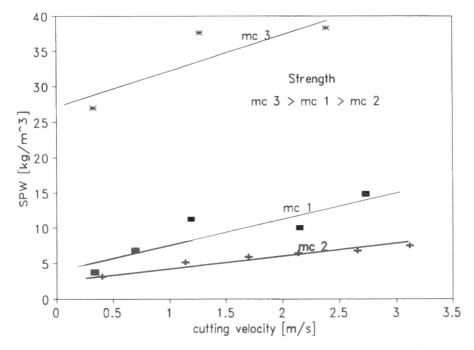

Figure 80 SPW [kg/m³] versus cutting velocity [m/s]. The influence of the mortar material strength
is shown

10.4.1.6 *The Vickers hardness of the dredge steel and wedge steel chisels*
Vickers hardness measurements have been made on the sections of these chisels at
increasing distance from the wear-flat after the experiments (Figure 75). Dredge steel
has a higher Vickers hardness near the wear-flat than wedge steel, but was more
susceptible to wear in the experiments. The lower hardness (brittleness) of wedge
steel favours microploughing as the abrasive wear mechanism over microcutting
(chapter 3). Microploughing is less severe compared to microcutting. Besides a lower
hardness allows rock fragments to be pressed into the steel. Rock fragments
embedded into the wear-flat of a chisel may protect the underlying steel against wear.
Pieces of rock embedded into the wear-flat have only been found in case of the
wedge steel chisel.

   Therefore, a higher Vickers hardness of the steel does not necessarily implicate a
higher wear resistance of a steel cutting tool.

10.4.2 *Experiments on different mortars*

Experiments with variation of the cutting velocity have been carried out on most
mortar compositions listed in Table 8. The feed was kept constant for all test runs
at approx. 0.65 mm/revolution. A critical velocity (V_{cr}) was not found for every
mortar composition in the range of the applied cutting velocities. The results of the

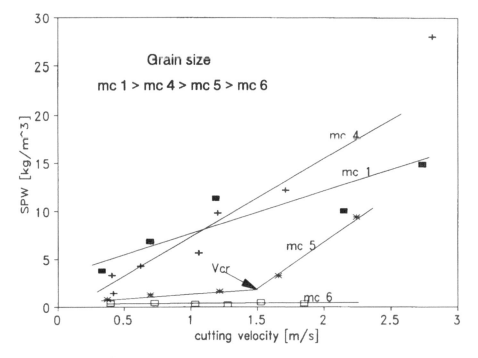

Figure 81 SPW [kg/m³] versus cutting velocity [m/s]. The influence of the grain size of the quartz grains is shown

experiments could be grouped into three types:

1. The rate of wear is constant for the full range of cutting velocities. Probably V_{cr} is higher than the maximum cutting velocity (3 m/s) (e.g. mc 6).

2. The rate of wear remains constant until the V_{cr} and increases with a further increase of the cutting velocity (e.g. mc 8 and mc 5).

3. The rate of wear increases from the lowest cutting velocity onwards. The V_{cr} is probably smaller than the smallest cutting velocity tested (0.3 m/s) (e.g. mc 1, mc 2, mc 3 and mc 4).

10.4.2.1 *Results of experiments on three mortars of different strength*
Mc 1, mc 2 and mc 3 contain 65 vol.% quartz with a grainsize of 1.5 mm and differ only in strength (UCS, BTS and Is50) (Strength mc3 > mc1 > mc2). Other properties like porosity and grain shape are approximately the same. In Figure 80 a graph of the SPW versus the cutting velocity shows the influence of the strength on the specific wear at different cutting velocities. An increase of the cutting velocity results in an increase of the SPW for all three mortars. In this range of cutting velocities a V_{cr} does not manifest itself. V_{cr} is probably smaller than or equal to 0.3 m/s. A higher strength of the mortar results in higher SPW values.

10.4.2.2 *Results of experiments on four mortars of different grain size*
Mc 1, mc 4, mc 5 and mc 6 contain 65 vol.% quartz and are approximately of the

same strength, however they differ in grain size (grain size mc 1 > mc 4 > mc 5 > mc 6). Other properties like porosity and grain shape are approximately the same. In Figure 81 a graph of the SPW is shown as a function of the cutting velocity for the four mortar compositions. A larger grain size generally causes a higher SPW value. The dependence of the SPW on the cutting velocity differs for the four types of mortar. Mc 6, which has the smallest grain size, does not show an increase of the SPW when the cutting velocity is increased. In case of mc 5, the SPW increases with increase of the cutting velocity only significantly at cutting velocities higher than 1.5 m/s and results of the experiments on mc 1 and mc 4 show an increase of the SPW with the cutting velocity for the complete range of cutting velocities tested. A V_{cr} is found only in case of mc 5 at about 1.5 m/s. In case of mc 6 the V_{cr} is probably higher than 3 m/s and in case of mc 1 and mc 4 it is probably lower than 0.3 m/s.

10.5 COMMENTS ON THE EXPERIMENTS ON ARTIFICIAL ROCK (MORTAR)

The properties of the mortar were considered to be virtually constant in time after a curing period of 28 days after casting. Actually, the chemical reactions in cement continue for more than a year, affecting the properties of a mortar. For example, the strength of a mortar increases in time (the maximum unconfined compressive strength of mortar is not supposed to deviate much from the unconfined compressive strength at 28 days after casting). Since not all experiments were carried out on mortars with the same curing periods, the properties of the mortars assessed after 28 days of curing period are not the same as the properties of the mortars during the experiments. However, the error is supposed to be small relative to the accuracy of the assessment of the mortar (rock) property parameters considered.

The wear resistance of the chisel decreases, because of high temperatures, which may develop at the wear-flat of a chisel during a scraping test experiment. However, if the wear path is short (5 m) the steel structure will be affected less than at longer wear paths (25 m). Most tests were performed with wear path lengths of at least 25 m. A higher rate of wear, expressed in gram of steel removed per meter sliding distance, was observed for a higher wear path length. The influence of the length of the wear path on the experimental results has not been taken into account, since it was supposed to be very small for most experiments. Only at large values of feed, mostly in wear mode III, the wear path is very short compared to the wear path in other experiments, due to the limitations of the test. However, an error in the rate of wear in wear mode III, although relatively large compared to the actual rate of wear, can be neglected since the rate of wear at these values of feed is very small compared to the rates of wear in wear mode I and II.

Attempts were made to measure the temperatures at the wear-flat of a chisel, during the tests. Unfortunately, thermocouple measurements failed, because in the vicinity of the wear-flat the thermocouples damaged.

Although the experiments were designed to keep the chisel shape constant, some experiments showed a change of the chisel shape (especially experiments on rock or mortar types with small grain sizes). Since the chisel shape may affect the chisel cutting characteristics, this may have biased the final results.

CHAPTER 11

Test results on natural rock types

11.1 INTRODUCTION

Natural rock differs in many aspects from the mortars used in the experiments described in chapter 10. However, many types of natural rock exist and some of them may show a similar behaviour with respect to rock cutting tool wear. Five types of sandstone and two types of limestone were characterized by the scraping test and the test results were compared to the results of the experiments on mortar.

11.2 EXPERIMENTAL RESULTS

Experiments were carried out on 5 types of sandstone and two types of limestone, the rock material properties of which are listed in table 9. Petrographic descriptions and photographs of thin sections of these rock types can be found in Appendix II. The experiments were carried out at feeds ranging from 0 to about 2 mm/revolution and at cutting velocities ranging from 0.4 to 2 m/s. During each test the feed and the cutting velocity were kept constant.

11.2.1 *Experiments on sandstones*

Experiments at different feeds and cutting velocities were carried out on 5 types of sandstone (Table 9). The wear modes observed in the scraping test experiments on artificial rock were also observed in the experiments on the sandstones. However, the sandstones showed a steeper gradient of the SPW versus feed graphs in wear mode I. Test results on the sandstones are shown in graphs (Figure 82, Figure 83 and Figure 84) of the rate of wear (wedge steel chisels) versus the feed (left graphs), SPW versus the feed and SPE versus the feed (right graphs). The feed was varied up to about 2 mm/revolution (in a single case even up to 2.5 mm/revolution) and the rate and type of wear at higher feeds could be classified in wear mode III. A higher cutting velocity resulted in a higher rate of wear, especially in wear mode I. In case of Felser sandstone, Bentheimer sandstone and Elb sandstone C in the range of small feeds the SPW was greater at higher cutting velocities. It was found to remain

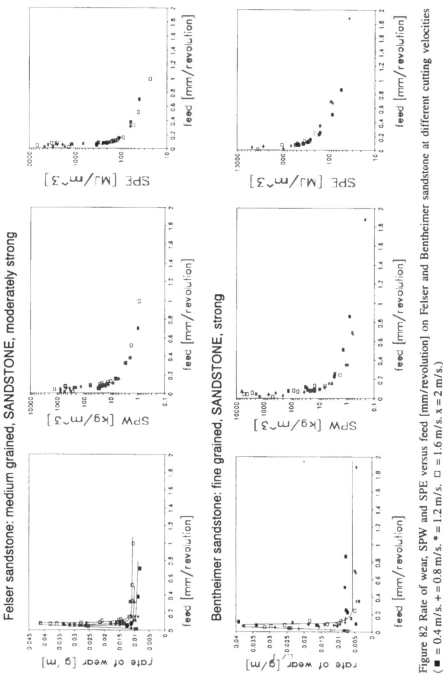

Figure 82 Rate of wear, SPW and SPE versus feed [mm/revolution] on Felser and Bentheimer sandstone at different cutting velocities
(■ = 0.4 m/s. + = 0.8 m/s. * = 1.2 m/s. □ = 1.6 m/s. x = 2 m/s.)

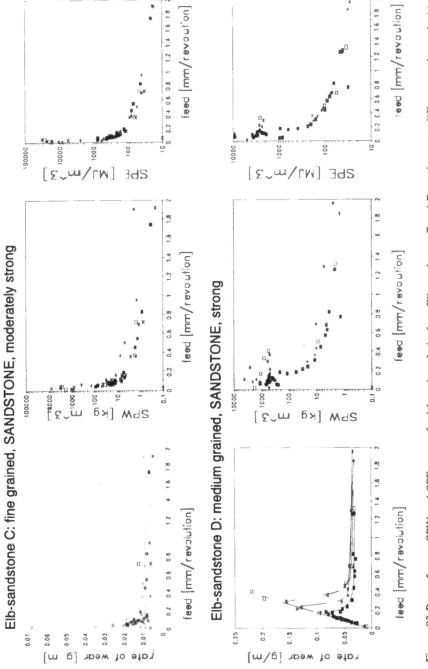

Figure 83 Rate of wear, SPW and SPE versus feed [mm/revolution] on Elb sandstone C and D sandstone at different cutting velocities
(■ = 0.4 m/s. + = 0.8 m/s. * = 1.2 m/s. □ = 1.6 m/s. x = 2 m/s.)

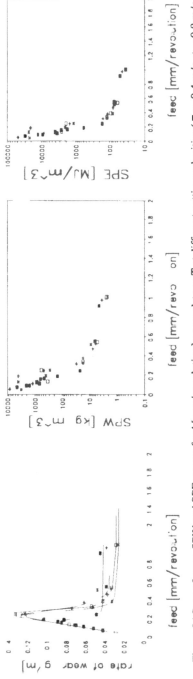

Figure 84 Rate of wear, SPW and SPE versus feed [mm/revolution] on sandstone T at different cutting velocities (■ = 0.4 m/s. + = 0.8 m/s. * = 1.2 m/s. □ = 1.6 m/s. x = 2 m/s.)

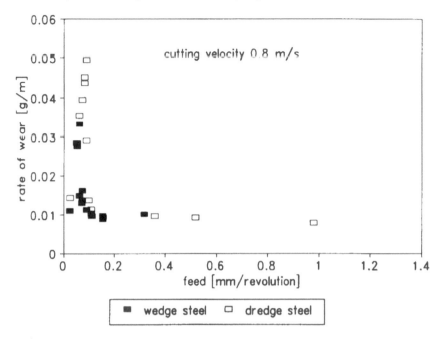

Figure 85 SPW [kg/m³] versus feed [mm/revolution] for wedge steel and dredge steel in experiments on Felser sandstone, tested at a cutting velocity of 0.8 m/s

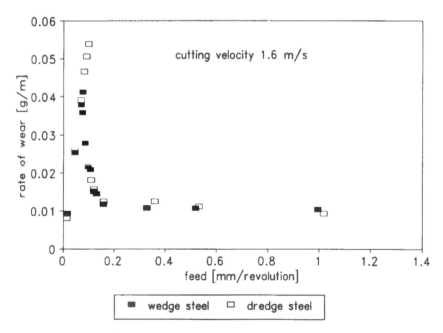

Figure 86 SPW [kg/m³] versus feed [mm/revolution] for wedge steel and dredge steel in experiments on Felser sandstone, tested at a cutting velocity of 1.6 m/s

approximately the same at different cutting velocities in the experiments on sandstone T. The feed at which the transition from one wear mode to another occurs shifts to higher feed values at higher cutting velocities. The results in the graphs concern only experiments with test chisels made of wedge steel. The work on mortars suggests that tests on dredge steel chisels would yield similar results. Slightly higher values of wear in wear mode I are expected in case of dredge steel chisels. In Figure 85 the rate of wear in experiments at a cutting velocity of 0.8 m/s on Felser sandstone is shown for dredge steel and wedge steel chisels. The rate of wear of the wedge steel chisels is a little bit lower in wear mode I. In Figure 86 the same trend can be seen. The only difference with Figure 85 is a cutting velocity of 1.6 m/s instead of 0.8 m/s.

11.2.2 *Experiments on limestones*

The types of wear, found in the experiments on limestones, which were composed mostly of calcite (Mohs hardness 3), only slightly resembled those found in the experiments on mortars and sandstones. No signs of scratches or grooves in the wear-flat of the chisel resulted from scraping or cutting the rock. The wear mode hypothesis, which is based on crushing of the abrasive grains, could not be applied to the limestones because of a lack of such grains in the rock. Still, some trends in the rate of wear, the cutting forces, the SPW or the SPE resembled those resulting from experiments on the mortars and sandstones. High temperatures, for example, occurred at the wear-flats of the test chisels in experiments at low feed values; and are probably the cause of the somewhat higher rates of wear at low feed values. The high temperatures are due to heat development by friction (rubbing) at low feed values where the chisel scrapes rather than cuts the rock. The type of wear can be classified as sliding wear (wear by soft abrasives, chapter 3). The rate of wear, the SPW and SPE of Sirieul and Euville limestone is shown in Figure 87. The cutting forces varied hardly over the tested feed range, which is also different in experiments on mortars and sandstones. Nevertheless the specific energy shows higher values at low feeds values due to a low production of cut rock material.

The cutting velocity affects the rate of wear, a lower cutting velocity causing a lower rate of wear. The SPW and SPE, however, seem to be less affected by the cutting velocity.

11.2.3 *Correlation of test results obtained in experiments on artificial or natural rock*

Expected specific wear values from Equation (6), based upon experiments on the artificial rocks (mortars), have been compared with the specific wear measured in experiments on natural rock types. Only experiments at a cutting velocity of 0.4 m/s were compared. In the calculations minerals other than quartz are added to the volume percentage of quartz if these minerals arc considered to be abrasive to the test chisels. Minerals with a Mohs hardness greater than the hardness of the steel chisel are considered abrasive, but only if they appear fresh. Feldspar for example is only added to the volume percentage of quartz if its weathering grade is fresh. Of course

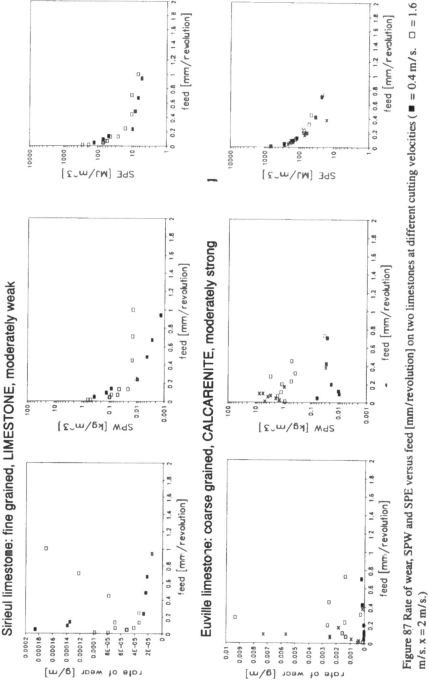

Figure 87 Rate of wear, SPW and SPE versus feed [mm/revolution] on two limestones at different cutting velocities (■ = 0.4 m/s. □ = 1.6 m/s. x = 2 m/s.)

the hardness of minerals and chisel steel are affected by the temperatures and stresses during the cutting process and this might affect the percentage of minerals in the rock which should be considered abrasive.

Equation (6) is only considered to be valid for rock types which have properties (grain size, volume percentage of abrasive minerals, strength) within the same ranges as the mortars which formed the basis of Equation (6). Only Felser sandstone is meets this requirement; the volume percentages of abrasive minerals of the other rock types are too high. In case of Felser sandstone, the fit of the predicted SPW values by Equation (6) with the experimental SPW values is reasonably good (correlation coefficient r^2 is 0.75). The predicted SPW values of the other sandstones only showed a reasonable fit with the experimental SPW values if the real volume percentage of abrasive minerals was substituted by a volume percentage of 60% (about the highest volume percentage in the tested mortars). Only the predicted SPW values of Bentheimer sandstone were a little high. The fit of the predicted SPW values with the experimental data improved even more when the coefficient of the feed of -5.5 was replaced by -7.5 which resulted in a steeper gradient of the first part of the curve. Equation (6) adjusted for the sandstones is given below (Equation (8)).

$$SPW = A*10^{-7.5feed} + B \qquad [\frac{kg}{m^3}]$$

$$(8)$$

in which

$A = 10^{-6.554 \ +0.054UCS \ +0.108vol.\%ab.min. \ +1.489\sqrt{\Phi}}$

and

$B = -0.026BTS \ +0.016vol.\%ab.min. \ +(0.0078*BTS*\Phi*vol.\%ab.min.)^2$

If the vol.%ab.min. > 60% than fill in 60%

The UCS and BTS are expressed in MPa and the grain size (Φ) in mm. Instead of the volume percentage of quartz, the volume percentage of abrasive minerals (vol.%.ab.min.) should be used in the Equations to stress the possible abrasive nature of hard minerals other than quartz. The volume percentages of the solid constituents together with the porosity make up 100%. If the volume percentage of abrasive minerals exceeds 60%, then 60% should be used as input for Equation (8).

Scraping test experiments on a feldspar containing mortar showed that the specific wear of wedge steel chisels was the same as that for a comparable quartz bearing mortar. Abrasive minerals softer than quartz (as well as abrasive minerals harder than quartz), but harder than the tool (test chisel) material, should therefore be taken into account when the SPW is calculated by Equation (8).

In Figure 88 the correlation between the SPW values, predicted with Equation (8) and the experimental SPW values is shown. The correlation is good for Felser sandstone, Elb sandstone C and sandstone T. In case of Bentheimer sandstone and Elb-sandstone D, the SPW predicted by Equation (8) is too high compared to the experimentally found SPW data. The correlation coefficient r^2 of the predicted SPW values with the experimental SPW values, disregarding the data on Bentheimer sandstone and Elb sandstone D, is 0.86. In Figure 87 the experimental data (dots) of the five tested sandstones obtained in the range of feed from 0.05 to 1.35 mm/revolution are shown together with the corresponding curves predicted with Equation (8).

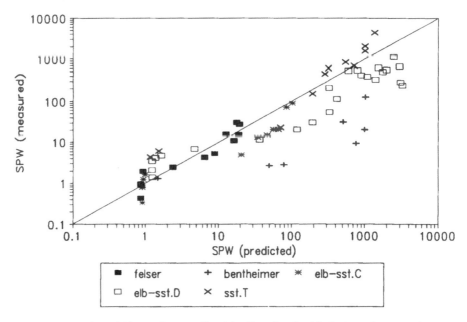

Figure 88 Correlation of SPW values predicted by Equation 8 with the experimental SPW values for the sandstones

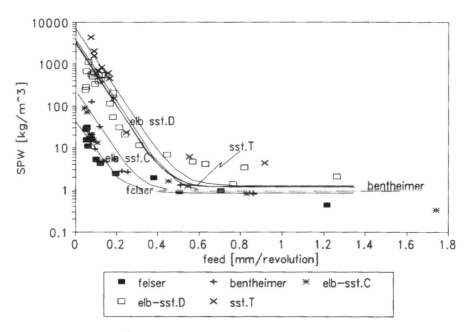

Figure 89 The SPW [kg/m^3] versus the feed [mm/revolution] for the sandstones. Experimental data are the dots. The lines represent the trend of SPW versus feed, predicted by Equation 8

11.3 COMMENTS ON THE EXPERIMENTS ON NATURAL ROCK

The type of cementation, texture and mineralogy of natural rock types differ mutually and from mortar. These differences may influence the rate and type of wear. However in case of the five sandstones, in general the same trends of rate and type of wear were observed in both the experiments on natural rock and artificial mortar rock. The two limestones caused a lower rate and a different type of wear. During testing of the sandstones and the mortars abrasive (grooving) wear dominated, whereas during testing of the limestones sliding wear prevailed (see chapter 3).

Predictions of the specific wear by Equation (6), which was based upon the experiments on mortars (chapter 10), were compared to the results of experiments on natural rock types. Only experiments at a cutting velocity of 0.4 m/s were compared. Differences between the specific wear predicted by this Equation and the specific wear observed in the experiments on natural rock types may reflect the differences in texture, cementation and mineralogy between mortar and natural rock or the difference in determination method of the parameters grain size and volume percentage of abrasive minerals. In the mortars these parameters were known before casting, but in case of the natural rocks they were determined by thin-section analysis. In appendix I the effect of different test methods on the porosity is shown as an example of the influence of the test method on the porosity values.

Equation (6), based on the mortar experiments, was adjusted to get a better fit with the experimental results on the sandstones. The adjustments were:

1. For the volume percentage of abrasive minerals, 60 percent should be used in the equation if the volume percentage of abrasive minerals is higher than 60 percent.

2. The coefficient of the feed should be -7.5 instead of -5.5, which results in a steeper gradient in the first part of the graph.

The influence of cutting velocity on the wear process is as yet not very clear and was therefore not included in the regression analysis.

CHAPTER 12

Relation of the scraping tests with rock cutting projects in practice

The scraping tests executed in this research programme are only remotely related to rock cutting projects in practice. Many factors, controlling wear and cutting processes in-situ, are not included in the small scale laboratory experiments. It is therefore questionable whether direct application of experimental results in practice is possible. In any case, the interpretation of the experimental results should be applied only in a qualitative manner, even then with the restriction that the dominating wear and cutting processes in the experiments should be the same as those in practice. Some aspects to be considered are for instance:

1. In the laboratory experiments the chisel continuously cuts or scrapes the rock (or mortar) while in dredging for example the chisel cuts the rock discontinuously: only for about 25% of the total working time.

2. In the experiments only wear due to the sliding action of a chisel over a rock surface is considered, while in dredging and trenching chisel impact may also play a role.

3. The experimental results apply to the materials involved and possibly also to materials with similar engineering properties. This means that the wear processes investigated and wear models derived may be applicable to steel but not to tungsten carbide and, besides that, only for steel tools cutting soft to medium strong granular rock types.

4. In the scraping tests the cutting velocity has been varied from 0.3 m/s up to 3 m/s and the feed from 0 up to 2.5 mm/revolution. Extrapolation of the test results to practice with cutting velocities and feed values beyond these ranges is questionable. In this respect it should be noted that in dredging cutting velocities fall in the range of 6-10 m/s; in trenching a usual range is 3-4 m/s. Feed values depend on rock characteristics but range usually from 0 to 10-30 mm per revolution per cutting tool.

5. The cutting experiments are performed in a dry environment, which may compare well to excavations on dry land, like trenching and ripping. However, in case of dredging, water may play an essential role in determining the type of wear processes.

Despite these shortcomings, an effort was made to compare the laboratory results to the wear, measured in rock cutting projects in practice. In dredging, trenching, ripping etc. chisels, bits or pick points, mounted on a cutter head, arm or chain often traverse the rock in an arc type of cutting path, thus producing a volume of rock with

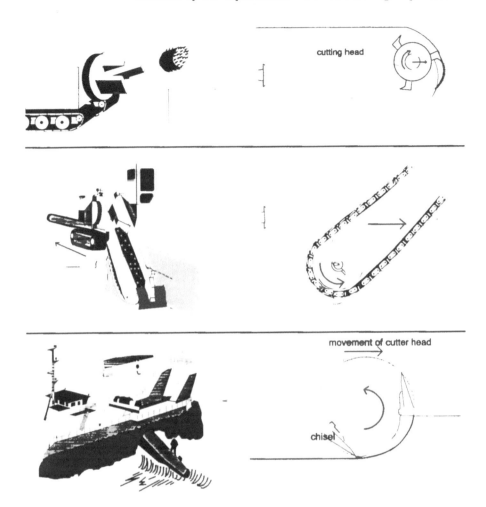

Figure 90 Artist impressions of a horizontal-turret boom-type tunnelling machine (after J.B.Edwards et al. 1993), A T-850 Vermeer trencher and a rotary cutter suction dredger with their cutting principles

increasing thickness towards the end or the start of the traverse. In Figure 90 some cutting paths of rock cutting tools, mounted on different types of rock cutting machines, are shown. In all cases the cutting principle involves a gradual increase of the depth of cut (or feed) of the cutting tool in the rock during a cut. (Only picks or chisels acting in the same plane of rotation are shown in the figure.) The increase of the cutting depth in practice is comparable to the variation of the displacement rate or feed of the chisels into the rock (or mortar) in the scraping test experiments.

In Figure 91 the effect of the increase of the feed of chisels mounted on a rock

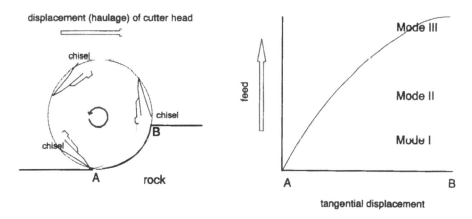

Figure 91 Rock cutting by a chisel mounted on a rock cutter head. Increase of feed from point A to point B may cause a change of dominating wear mode

cutter head acting in one plane of rotation[29] on the wear mode is shown. The chisel starts in point A in mode I. As the feed increases, when the chisel is moved from point A to B, the wear and cutting mode may change to mode II and finally to mode III. The boundaries between the ranges of feed, where different wear modes prevail may vary with rock properties.

Not only the displacement from point A to point B affects the feed. Pick-point interaction, for example, or blunting of the pick-points may decrease the feed. Part of the intact rock to be cut may already have been removed by another chisel. The resulting decrease of the feed of the chisel may cause a change of wear mode. Blunted pick-points slow down the haul velocity and therefore reduce the feed of a pick-point per revolution of the cutterhead.

In the scraping test experiments the feed was varied in different test runs resulting in various graphs such as the specific wear (SPW) or specific energy (SPE) versus the feed and the rate of wear versus the feed. These graphs in combination with operating parameters of the cutting machine, such as advance rate and cutting velocity, might indicate which wear mode is likely to prevail in practice. In Figure 92 the change of wear mode prevalence with variation of the feed of a single chisel is illustrated.

If in practice cutting and wear processes are unfavourable, they may be changed, for example by changing the cutting velocity or the advance rate or by a better tuning of the cutting device by changing the design of the cutter head. In this respect the results of the scraping test may give valuable guidance. For example, a rock cutting suction trailing dredger performed better from a wear-point of view than a rotary rock cutter suction dredger in the same rock formation near Australia (Prof. de

---

[29] The feed of a pick-point mounted on a rock cutter head is related to the haul velocity of the cutter head and inversely related to the rotation velocity of the cutter head and the number of blades of the cutter head. Only the number of cutting blades with the same pick-point positions should be considered.

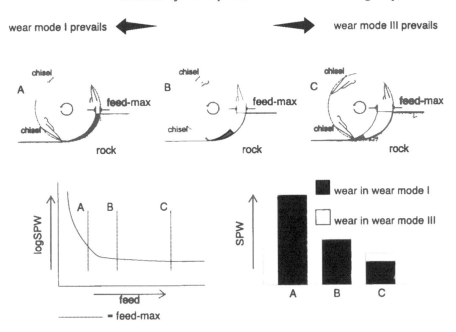

Figure 92 The effect of the feed on the specific wear (SPW) of a rotary cutting chisel. Wear mode I contributes most to the total SPW. The average SPW decreases rapidly if the maximum feed (feed-max) increases

Koning[30] pers.comm.). The better performance of the trailing cutter dredger (pick-point consumption was 20 times higher in case of the rotary cutter suction dredger) can be explained by looking at the scraping test experiment results. The unfavourable wear mode I, observed in the wear experiments at small feeds, is only likely to occur in case of the rotary cutter dredger because of its arc shaped cutting path starting each cut with a small feed (in case of undercutting; in case of overcutting small feeds occur at the end of each cut). Another example of application of the test results to practice may be to tune the spacings of the bits on a cutter head or chain with respect to the grain size of the rock to be cut. Smaller spacings of the bits may be necessary to achieve a favourable interaction between the bits, if rock types with smaller grain sizes are to be cut. Next to improvements of the cutting machinery, cutting conditions etc., a better insight in the wear and cutting processes may also lead to a better prediction of the expected amount of wear. The average specific wear is found by integrating the specific wear over the complete feed range and dividing it by the maximum feed reached. The lower limit of the integration is always zero and the upper limit is the maximum feed (feed_{max}), which is a function of the advance rate, the rotation velocity and the number of chisels acting in one plane of rotation.

---

[30] Professor de Koning has worked in the dredging industry and carried out research related to dredging at the Faculty of Mechanical Engineering of Delft University of Technology.

At the end of 1993 a study was carried out on rock cutting trenchers (T-850 trenchers manufactured by Vermeer Manufacturing Company), working in four different types of rock (M. Giezen 1993). The effect of changing rock types on tool wear and production rate of these trenchers was studied. Next to the rates of tool wear, also the type of wear process was observed and related to the wear modes as described in this thesis. Two of the trenchers worked in quartz containing, granular rock types (a granite and a sandstone) and wear due to sliding of the bits against the rock seemed to be the main cause of wear. These conditions make a comparison between these trenching cases and the laboratory test results possible.

12.1 DESCRIPTION OF THE T-850 ROCK CUTTING TRENCHER

A trencher resembles a chain saw, which cuts or saws through rock or soil. The chain is covered by baseplates on which bits are located in V-shaped patterns. The chain rotates and carbide-tipped bits cut through the rock while the trencher device itself moves backwards.

In principle the trencher always operates at full power (fuel consumption is at its maximum), when cutting through hard rock. The total power is roughly divided into the energy, needed to obtain some travel speed of the trencher (the ground drive), and the energy, needed to keep the digging chain going round. The digging chain speed remains constant (about 3.3 m/s). The travel speed or advance rate is thus adapted to the energy needed to cut a particular rock mass (0 to 27 m/min).

The trencher can be operated manually and automatically. The "auto creep", a sort of automatic pilot, controls the division of power (or energy). If the trencher operates at auto creep, the operator has virtually no influence on the rock cutting process.

12.2 CALCULATION OF THE FEED OF A SINGLE BIT OF THE T-850 ROCK CUTTING TRENCHER

In Figure 90 the cutting principle of a trencher is shown and the cut of a single bit visualized. The feed of a bit gradually increases as it moves along the arc type of cutting path from zero to a maximum feed value. Then the cut becomes straight until the bit leaves the rock. The maximum feed of a single bit of a rock cutting trencher (the feed at the end of the arc-shaped cut) can be calculated from the rotational frequency (rpm) of the chain, the travel speed of the trencher, the number of bits on the chain acting in the same vertical plane and the angle of the boom with the earth surface (Equation (9)). The maximum feed of a single bit was calculated by Equation (9) for a trencher which worked in a granite (in Schnarrtanne, Germany) and a sandstone (in Oberndorf, Germany)[31]. The angle of the boom with the earth surface

---

[31] According to BS 5930:1981 the Scharrtanne granite can be described as a pinkish white, coarse grained, wholly discoloured, mineralogically altered, micro-fractured GRANITE, moderately strong.
The Oberndorfer sandstone can be described as a yellowish grey, medium grained, partially discoloured, calcareous SANDSTONE, moderately strong.

$$feed_{max,trencher} = \frac{travel\ speed\ (mm/h)}{rpm_{chain} * 60\ (min) * no.\ of\ bits\ acting\ in\ the\ same\ vertical\ plane} * sin\alpha$$

expressed in [*mm per revolution of the chain per bit*]

(9)

α = *angle of the boom with the earth surface*

deducted from the trench depth to be about 52°. The average maximum feed of the trencher working in the granite and the sandstone is respectively 0.80 mm and 1.97 mm per revolution of the rock cutting chain per bit. If the travel speed of the trencher is not known, which is often the case before the start of a trenching project, a prediction of the expected travel speed should be estimated in order to obtain the maximum feed values.

The travel speed of a rock cutting trencher depends upon the applied machine power, the width of the trench, the thickness of the solid rock volume to be trenched and the excavation properties of the rock material. Rock mass properties are not considered since these are often describing rock behaviour on a larger scale than the scale of the thickness of a cut of a bit of a trencher, which is in the order of millimetres to centimetres (M.Giezen 1993). Especially the BTS seems to be promising as a descriptor of the cuttability of rock, since it shows a linear relationship with the cutting force in the scraping test experiments (Deketh 1993). Thus for a specific trencher a relationship may be found between the feed$_{max}$ and the BTS, trench width and solid rock thickness. The maximum feed for a T-850 trencher is related to these parameters in Equation (10). Equation (10) is based on the Schnarrtanne and Oberndorf project. The coefficient of 2.6 MNmm is found by matching the maximum feed values, measured in-situ, with the parameters in the Equation.

$$feed_{max,T850\ trencher} = \frac{2.6\ (MNmm)}{BTS\ (Mpa) * trench\ width\ (m) * solid\ rock\ thickness\ (m)}$$

(10)

expressed in [*mm per revolution of the chain per bit*]

(From the maximum feed also the travel speed (advance rate) or the production of a T-850 trencher can be deducted).

12.3 THE SPECIFIC WEAR IN THE SCHNARRTANNE GRANITE AND THE OBERDORFER SANDSTONE, PREDICTED ON THE BASIS OF RESULTS FROM THE SCRAPING TEST EXPERIMENTS

In Table 16 some properties of the two rock types are listed. The complete petrographic description can be found in Appendix II.

Applying Equation (8) to the Schnartanne granite and the Oberndorfer sandstone results in the graphs shown in Figure 93 and Figure 94 respectively. The graphs are

Table 16. Properties of the granite from Schnarrtanne and the sandstone from Oberndorf. The values in brackets in the column of the volume percentage of abrasive minerals include all the minerals which have a possible abrasive nature, whereas the values above are the volume percentage of quartz only. In both volume percentages the porosity is taken into account. * This value is based on the assumption that the microcracks act as grain boundaries of the weathered quartz grains. ** This value disregards the possible effect of microcracks

Rock name	Grain size (mm)	Vol. % abrasive minerals (%)	Grain shape	UCS (MPa)	BTS (MPa)	Is50 (MPa)	E-mo-dulus (GPa)	Dry density (Mg/m³)	Poro-sity (%)
gra-nite Sch.	0.7* 2.1**	29 (88)	an-gular	37	2.7	1.4	11	2.5	5
sand-stone Ober.	0.56	35 (49)	sub-roun-ded	47	2.3	2.2	15	2.2	19

supposed to be fingerprints of the abrasive capacity of these rock types for a range of feeds. The ranges of feed of a single bit of the trencher are known for both rock types; they are marked in the graphs. In case of the Oberndorfer sandstone the SPW versus feed graph could be produced straightforward from the measured rock parameters. In case of the Schnarrtanne granite it is not very clear how to interpret the effect of the weathered minerals (feldspars) on wear. Equation (8) does not take mineral weathering into account. Weathered feldspars are considered to be less abrasive if not non-abrasive. Moreover , the grain size of the quartz grains can be determined with or without taking the micro-cracks in the quartz grains into account. If the micro-cracks are considered as grain boundaries the grain size of the quartz grains is considerably smaller (see Table 16). In Figure 93 the SPW versus feed is shown for the Schnarrtanne granite for the two differently determined grain sizes and the two differently determined volume percentages of abrasive minerals. Obviously a different input in Equation (8), due to a different interpretation of the thin section analysis, results in a considerable difference of the SPW versus feed graph. Hence the prediction of the specific wear by Equation (8) is not accurate in case of rock

$$SPW_{av.} = \frac{A \int_{0.05}^{feed_{max}} (10^{-7.5feed} + B)\, d(feed)}{feed_{max} - 0.05} = \frac{-\dfrac{A}{7.5 ln10} \left[10^{-7.5feed}\right]_{0.05}^{feed_{max}} + B\left[feed\right]_{0.05}^{feed_{max}}}{feed_{max} - 0.05}$$

$SPW_{av.}$ is expressed in $\dfrac{kg}{m^3}$

$$A = 10^{-6.554\ +0.054UCS\ +0.108vol.\%ab.min.\ +1.489\sqrt{\phi}}$$

$$B = -0.026BTS +0.016vol.\%ab.min. +(0.0078*BTS*\Phi*vol.\%ab.min.)^2$$

if the vol.% of abrasive minerals > 60% than fill in 60%

(11)

Table 17. The predicted values and the field measurements of specific wear of the bits of the T-850 trenchers working in the Schnarrtanne granite and the Oberndorfer sandstone. The values with an asterix (*) are specific wear values of all the bits, the other values of a single bit. The grain size and, respectively, the volume percentage of abrasive minerals, which were used as input for Equation (11), are in case of [1] 2.1 mm and 60 %, in case of [2] 0.7 mm and 60 %, in case of [3] 2.1 mm and 29.3 % and in case of [4] 0.7 mm and 29.3 %

SPW$_{av.}$			
Schnarrtanne granite		Oberndorfer sandstone	
predicted by equation (11) [kg/m^3]	field measurement [g/m^3]	predicted by equation (11) [kg/m^3]	field measurement [g/m^3]
405[1]			
50[2]	36.1*		35.3*
2.3[3]	0.21	3.7	0.25
0.6[4]			

types which bear weathered minerals. If Equation (8) is integrated over the feed range of a rock, cut during trenching, and the result is divided by the feed range, the average SPW value (SPW$_{av.}$) is obtained. This is a measure for the total specific wear of a bit during trenching. This results in Equation (11). For feed values smaller than 0.05 mm per revolution, Equation (11) is considered to yield SPW values, which are too high. In experiments on some mortars at feed values close to zero, the SPW decreased as the feed decreased, whereas the SPW, calculated with Equation (11), increases with decreasing feeds.

In Equation (11) the feed is considered to increase linearly from 0.05 mm/revolution at the start of the cut to the maximum calculated feed value at the end of the arc-shaped cut.

 In reality the increase of feed is larger at the start of a cut than at the end. In case of the trencher, at the end of the arc-shaped part of the cut, the bit continues to cut a constant depth until it leaves the rock or soil at the surface. The part of the cut at constant depth of cut is disregarded in the calculations, because in this part the feed is relatively high and therefore the rate of wear is low and can be neglected. Moreover, in this part the bits often cut through weathered rock or soil, which further mitigates the rate of wear. Another difference between the predicted SPW and the amount of wear occurring in practice, may be caused by the difference in cutting velocity in the experiments and that in practice.

12.4 COMPARISON OF THE FIELD RESULTS WITH THE EXPERIMENTAL RESULTS

In Table 17 the specific wear (SPW$_{av.}$) for the Oberndorfer and Schnarrtanne trenching project predicted with Equation (11), are compared with each other and with the measured specific wear values in the field. Since it was not clear how much

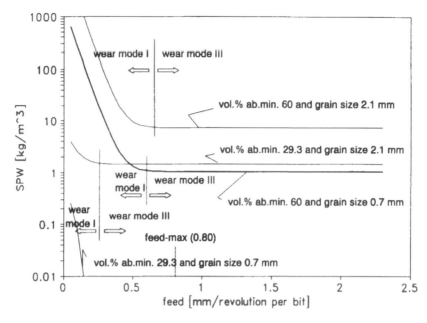

Figure 93 Predicted specific wear in the range of feed at which the T-850 trencher was working (from 0 to feed-max) in the Schnarrtanne granite. Wear modes are separated by vertical lines

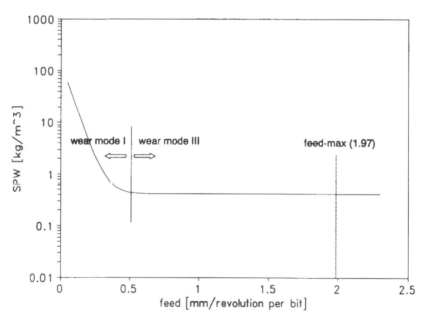

Figure 94 Predicted specific wear in the range of feed at which the T-850 trencher was working (from 0 to feed-max) in the Oberndorfer sandstone. Wear modes are separated by vertical lines

Figure 95 Part of a bit. Irregularly divided grooves in the steel, partly infilled, may point out a three-body wear process. Top of the bit at the left side of the photograph; a (small) scale-unit equals 1 mm

Figure 96 Part of a bit. Relatively long clean continuous grooves in the steel may point out a two-body wear process. Top of the bit at the left side of the photograph. A (small) scale-unit equals 1 mm

the weathering and microfracturing of the minerals in the granite affected the wear processes, four different possibilities of grain size and volume percentage of abrasive minerals were used to calculate the predicted specific wear with equation (11). The specific wear measured in the field is about the same in both projects. If the comparison is made with the wear of one bit of the trencher, the predicted specific wear determined by Equation (11) and differs a factor of about 10000 from the specific wear recorded in the field. If the wear of all the bits per m³ of trenched rock (total number of bits on the trenchers is 141 for Oberndorf and 172 for Scharrtanne) is compared with the predicted average specific wear by Equation (11), the specific wear of the bits in the field is about a factor 100 lower than that calculated with Equation (11). The SPW_{av} per bit indicates how often each bit should be replaced over a certain trench volume. The SPW_{av} of all the bits takes differences in number of bits on different trenchers into account. The SPW_{av} values, calculated with Equation (11), are therefore not absolute values, but only suitable for relative comparison between the trenching projects. The range of predicted SPW_{av} of Schnarrtanne is too large to make a sensible comparison of the predicted SPW_{av} between both projects. Clearly, (small) differences in rock property parameters have great effect on the SPW_{av} (These parameters are implemented in the exponent in Equation (11)). A correct analysis of thin-sections and the interpretation of this analysis is therefore of outmost importance. However at this stage it is not clear how this should be done.

The type of wear might be predicted by the SPW versus feed graphs in Figure 93 and in Figure 94. As can be seen in the graphs, the SPW is high at low feed values and decreases rapidly with increase of the feed until it reaches a more or less constant low level at larger values of the feed. The feed value at which the SPW reaches this constant level is supposed to be the point at which wear mode III begins to dominate in the wear process. Comparison of the magnitudes of the feed ranges before and beyond this feed value, indicates how much each wear mode is expected to contribute to the total wear in the working range of feeds of the trenchers. In the case of Schnarrtanne, wear mode I contributes for about 31% to 75% (average 53 %) to the total wear. In the case of Oberndorf wear mode I is expected to contribute only for about 28% to the total wear process. (Wear mode II is not shown in the graphs because it is a mix of wear mode I and III and is therefore difficult to delineate).

Photographs of the surface of the bits were taken (Figure 96 and Figure 95). The observed wear phenomena agree with what was expected from the SPW versus feed graph interpretation. In case of the granite, wear mode I is expected to dominate. In fact clean grooves in the steel can be seen on the photograph, which indicates two-body wear and thus wear mode I. The irregularly divided grooves, partly filled by rock crushings, can be seen in the steel surface of a bit which worked in the sandstone. This points to three-body wear, which is the mechanism behind wear mode III, which, on the basis of the SPW versus feed graph was expected to be dominant.

The experimentally based results agree well with the type of wear observed in both trenching projects. However for a proper comparison of predicted and measured bit wear more study should be carried out on the interpretation of thin-sections to obtain correct values of the used rock property parameters.

CHAPTER 13

Conclusions

From a wear-point of view the test results showed that it is disadvantageous to cut rock at shallow cuts with limited penetration (small feed). By changing cutter head design (less blades), by changing operating control parameters (rotation velocity and haul velocity of the cutterhead), by increasing the machine power (as long as the machine parts do not fail due to the consequently higher forces) or by changing the operating principle (e.g. a trailing cutter head instead of a revolving cutter head) an increase of the feed of the cutting tools and thus less wear can be achieved.

The laboratory experiments prooved helpful in understanding the cutting and wear processes at small feed ranges. The effects of feed and cutting velocity have been investigated on 17 types of different artificial rocks (mortars), five types of sandstone and two types of limestone in more than thousand laboratory experiments. The most important observations in these experiments are listed below:

1. The increase of the feed of a chisel into the rock can lead to a change of wear mode (a change of the rate and type of wear). This change of wear mode goes together with a change of cutting mode. At low values of feed wear often takes place as a two-body wear process causing high temperatures at the wear-flat of the tool, which weaken the tool steel (wear mode I). The rate of wear is relatively high and the production of cut rock material is relatively low; the chisel scrapes the rock (cutting mode I). This results in relatively high values of specific wear.

   At high values of feed three-body wear is the main wear mechanism and lower temperatures occur at the wear flat (wear mode III). A relatively low rate of wear results. The production of cut rock material is higher; the chisel cuts the rock (cutting mode III). This results in relatively low values of specific wear.

   The transition from wear (and cutting) mode I to III is gradual and called wear (and cutting) mode II.

2. Experiments on artificial rock types (mortar) showed that the unconfined compressive strength, the Brazilian tensile strength, the grain size and the volume percentage of abrasive minerals affect the rate and type of wear and the specific wear. Increasing values of these properties cause an increase of the rate of wear, especially in the range of feeds at which wear modes I and II take place and they cause a shift in the feed values, at which wear modes change (from I to II and from II to III), towards larger values of feed. The shape of the abrasive mineral grains

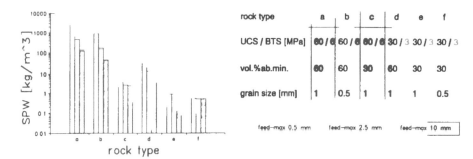

Figure 97 Differences in rock material properties and the maximum feed of a cutting tool causes differences in the expected specific wear. The rock types are theoretical

affects the rate of wear only in wear mode I. Mortars containing feldspar and mortars containing quartz cause the same scraping test experiment results. In Figure 97 the effect of differences in the above mentioned rock property parameters and the maximum feed of a cutting tool on the specific wear as expected from the experiments.

3. Natural sandstones display the same wear modes as the artificial rock types. Natural limestones cause sliding wear (wear by soft abrasives), resulting in elastic and plastic deformation, adhesion and surface fatigue of the steel at the wear-flat of the rock cutting chisel.

4. In general the cuttability of different rock types expressed as the specific energy, needed to cut these rock types, is affected by the same parameters and in the same way as the specific wear. Only the grain size does not affect the specific energy in the range of feeds at which wear mode III dominates.

5. The cutting force correlates better with the Brazilian tensile strength than with the unconfined compressive strength, at least in cutting (wear) mode III, which implicates that chip formation probably takes place by tensile failure of the rock.

6. The cuttability is affected by free surfaces. Cracks originating at a tool tip may run towards these surfaces, which may be created by cutting tools travelling at either side of the tool during rock cutting practice. In rock types with smaller grain sizes, the propagation of the cracks towards the free surfaces becomes less. As a result the interaction between the cutting tools becomes smaller, what may result in an increase in the specific energy.

7. An increase of the cutting velocity causes an increase in the rate of wear, mainly in wear modes I and II. The disadvantageous wear mode I still dominates at higher feed values for higher cutting velocities. Experiments, performed at the same feed, with the cutting velocity as variable, show thus a critical velocity at which the rate of wear increases rapidly with a further increase of the cutting velocity. An increase of the rock material strength or an increase of the grain size of the abrasive minerals causes an increase of the specific wear at the present range of cutting velocities (0.3-3 m/s). The critical cutting velocity is affected by the grain size. A larger grain size causes a decrease of the critical cutting velocity.

8. Wedge steel chisels (steel Fe60K) performed better than dredge steel chisels (SRO 57N) in wear mode I. Dredge steel performed slightly better in wear mode

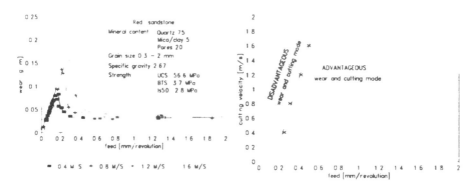

Figure 98 Rate of chisel wear in scraping tests on a sandstone as a function of the feed for different cutting velocities. At a higher cutting velocity a higher feed is needed for an advantageous mode of wear

III. The worse performance of dredge steel may be due to the sensitivity of dredge steel to steel weakening by the high temperatures, occurring in wear mode I. The initial hardness of the dredge steel drops whereas wedge steel seems to be self-hardening under the stresses induced by the cutting process.

Besides rock fragments, bedded into the wear-flat, were only observed in case of wedge steel chisels. These rock fragments may protect the wedge steel from further wear by the rock asperities.

Another cause may be the higher brittleness of dredge steel, resulting in a transition from microcutting to microploughing as the abrasive wear mechanism (microcutting causes more wear than microploughing).

CHAPTER 14

Closing remarks and recommendations

The improvement of the interpretation of thin-section analysis is considered as the next step in the research of wear of rock cutting tools to get hold of the factors controlling wear. Thin-sections can easily be interpreted differently, because the interpretation is subjective, and because the criteria of the determination of petrographic parameters, which are considered to be relevant to wear, are not clear at this stage.

Wear is a system dependent process and therefore many parameters of different nature and origin affect the processes of wear. In this research some of those parameters, considered to be important to wear, have been investigated, but many others still remain to be investigated. In practice, for each project different parameters may be of importance and therefore a thorough analysis should be carried out before a (quantitative) estimation of the expected rate and type of wear can be made.

Nomenclature

Feed | radial displacement the support of the lathe per revolution of the turning head of the lathe [mm/revolution] or the displacement of a bit,pick or chisel into the rock perpendicular to the direction of rotation of the cutter head or chain of a rotary cutting machine.

Cutting depth | thickness of the rock-cut per revolution of turning head of the lathe [mm/revolution]
thickness of the rock-cut of a chisel mounted on a rock cutter head

Cutting velocity | velocity of the chisel relative to rock at the chisel-rock contact in the direction of cut [m/s]

Cutting forces | forces acting on chisel during cutting or scraping of rock

Cutting force | force acting on chisel parallel to direction of cut [N] (also shear or tangential force)

Normal force | force acting on chisel perpendicular to the rock surface [N]

Lateral force | force acting parallel to the rock surface and perpendicular to direction of cut [N]

Wear rate | difference in weight of a chisel before and after an experiment divided by the length of cut [g/m]

Specific wear | difference in weight of a chisel before and after a specific cutting experiment or project divided by the volume of rock cut [kg/m^3]

Cutting mode ratio | ratio of cutting depth and feed

Specific energy | energy needed to cut one cubic meter of rock specific to a specific cutting apparatus and to the circumstances of cutting [MJ/m^3]

List of abbreviations and symbols

UCS	Unconfined Compressive Strength
BTS	Brazilian Tensile Strength
Is50	Point Load Strength Index
RTI	Rock Toughness Index
SPW	Specific Wear
SPE	Specific Energy
vol.%ab.min.	Volume Percentage of Abrasive Minerals
F_x	Average Lateral Force
F_y	Average Normal Force
F_z	Average Cutting Force
Φ	Grain Size
$feed_{max}$	Maximum Value of Feed Reached in a Rock cut
SPW_{av}	Specific Wear Averaged over a Rock Cut

References

Abdullatif, O.M. & D.M.Cruden 1983. The relationship between rock mass quality and ease of excavation. *Bull Int. Assoc. Eng. Geol.*: vol. 28, 183-187.

Bell, F.G. 1992. *Engineering in Rock Masses*. Butterworth, Oxford.

Bisschop, F. 1991. The analysis of a laboratory cutting and abrasion test to be applied in rock cutting dredging. Memoirs of the Centre for Engineering Geology in the Netherlands, no. 81.

Bray, R.N. 1979. *Dredging: A Handbook for Engineers*. Edward Arnold, London.

Braybrooke, J.C. 1988. The State of the Art of Rock Cuttability and Rippability Prediction, *Proc. Fifth Australia-New Zealand Conf. on Geomechanics Sydney*: 13-42.

Clark, G.B. 1987. *Principles of Rock Fragmentation*. John Wiley & Sons, New York.

Cools, P.M.C.B.M. 1993. Temperature Measurements Upon the Chisel Surface During Rock Cutting. *Int.J.of Rock Mech. and Min.Sc. & Geomech. Abstr. vol.30, No1*: 25-35.

Dalziel, J.A. & A.Davies 1964. Initiation of Cracks in Coal Specimens by Blunted Wedges. *The Engineer, Jan. vol. 31*.

Deketh, H.J.R. 1991. Determination of optimum conditions for the modified pin-on-disc test. Memoirs of the Centre for Engineering Geology in the Netherlands, no 85, Technical University Delft, Faculty of Mining and Petroleum Engineering.

Deketh,H.J.R. 1993. Rock properties relevant for the assessment of abrasiveness and excavatability of rock, applied to dredging. *Proceedings of the CEDA-Dredging Days, Experience of the Application of Research Results to Dredging Practice, Amsterdam*, Central Dredging Association.

Davids, S.W. & P. Adrichem 1990. Het testen van een classificatie-apparaat voor de slijtage van tanden van snijkoppen bij het snijden van rots. Rapport nr. 90.3.GV.2671, Technische Universiteit Delft, Faculteit Werktuigbouwkunde en Maritieme Techniek.

Dubugnon, O. & P. Barendsen. JAARTAL Small Scale Model Testing, A New Approach in TBM Development. BOEK Institut CERAC SA, 1024 Ecublens, Switzerland.

Edwards, J.B., R.J. Hollands, S.Yaacob & S.M. Hargrave 1993. Robotics for the control of boom-type tunnelling machines for soft- and hard rock application. *Mine*

Mechanization and Automation. Proceedings on the second international symposium on Mine Mechanization and Automation, Luleå Sweden: 221-230 Balkema, Rotterdam.

Evans, I. 1974. Relative Efficiency of picks and discs for cutting rock. *Proc. of 3rd Congress ISRM, Vol. II.B Denver, USA*: 1399-1406.

Evans, I. 1962. A Theory of the Basic Mechanics of Coal Ploughing. *Proc. of the International Symposium on Mining Research, University of Missouri, vol II*: p. 761.

Ewendt, G. 1989. Erfassung der Gesteinsabrasivität und Prognose des Werkzeugversleisses beim maschinellen Tunnelvortrieb mit Diskenmeisseln. Bochumer geologische und geotechnische Arbeiten, Heft 33, Institüt für Geologie, Ruhr-Universität Bochum.

Farmer, I. 1983. *Engineering Behaviour of Rocks*. Second edition Chapman & Hall Ltd. London.

Fowell, R.J. 1970. A simple method for assessing the machinability of rocks. *Tunnels and Tunnelling*.

Fowell, R.J. 1991. Cuttability assessment applied to drag tool tunnelling machines. *Proc. 7th Int. Congress of the International Society for Rock Mechanics, Vol 2, Aachen*: 985-990 Balkema, Rotterdam.

Franklin, J.A., E. Broch & G. Walton 1971. Logging the mechanical behaviour of rock. *Trans. Inst. Metall., vol.80*: section A-Mining industry A1-9

Ford. L.M. & M. Friedman 1983. Optimisation of Rock-Cutting Tools Used in Coal Mining. *24th US Symposium on Rock Mechanics*: 725-731 Association of Engineering Geologists, Lawrence.

Friedman,M. & L.M. Ford 1983. Analysis of Rock Deformation and Fractures Induced by Rock Cutting Tools Used in Coal Mining. *24th US Symposium on Rock Mechanics*.: 713-723 Association of Engineering Geologists, Lawrence.

Gehring, K. 1987 Rock testing procedures at Va's geotechnical laboratory in Zeltweg. Internal report TZU 41. Voest Alpine Zeltweg, Austria.

Giezen, M. 1993 Rock Properties Relevant for Tool Wear and Production of Rock Cutting Trenchers. Memoirs of the Centre for Engineering Geology in the Netherlands, no 110, Technical University Delft, Faculty of Mining and Petroleum Engineering.

Handayan, J.M., E.R. Danek, R.A.D. Andrea & J.D. Sage,J.D. 1990. The Role of Tension in Failure of Jointed Rock. *Proc. of the Int. Symp. on Rock Joints, Loen Norway*: 195-202, Balkema, Rotterdam.

Hettema, M.H.H., C.J. de Pater & K-H.A.A. Wolf 1991. Effects of temperature and pore water on creep of sandstone. *Rock Mechanics as a Multidisciplinary Science, Proc. of the 32nd U.S. Symposium*: 393-404 Balkema, Rotterdam.

Hignett, H.J., 1984. The current state of the art of rock cutting and dredging. Miscellaneous paper GL-84-17, Department of the Army, US Army Corps of Engineers Washington, DC 20314-1000.

Hoogenbrugge, W.J., 1980. The influence of discontinuities on the excavation of rock by disc bottom cutter dredger. Memoirs of the Centre for Engineering Geology in the Netherlands, no 03, Technical University Delft, Faculty of Mining and Petroleum Engineering.

Hopkins, D.L., N.G.W. Cook & L.R. Myer 1990. Normal Joint Stiffness as A

Function of Spatial Geometry and Surface Roughness. *Proc. of the Int. Symp. on Rock Joints, pp 203-210, Loen Norway*: 203-210 Balkema, Rotterdam.

Iverson, N.R. 1991. Morphology of glacial striae: Implications for abrasion of glacier beds and fault surfaces. *Geological Society of America Bulletin, Vol.103*: 1308-1316.

Jager, W. 1988. An investigation on the abrasive capacity of rock. Memoirs for the centre of Engineering Geology in the Netherlands, No 52, Technical University Delft, Faculty of Mining and Petroleum Engineering.

Kenny,P. & S.N. Johnson 1976. An investigation of the abrasive wear of mineral cutting tools. *Wear Vol.36*: 337-361

Kirsten, H.A.D 1982. A classification system for excavation in natural materials. *The Civil Engineer in South Africa*, vol. 24, 293-306.

Koert, J.P. 1981. Trillend snijden van gesteenten. rapport no. CO/81/110, Technische Universiteit Delft, Faculteit der Werktuigkunde en Maritieme Techniek, Vakgroep Transporttechnologie, Delft.

Korevaar, B.M. & B. Pennekamp 1984. Staal en Gietijzer. Lecture Notes Mt7a, Faculty of Metal Sciences, Delft University of Technology.

Kutter,H.K. & H.P.Sanio 1982. Comparative study of performance of new and worn disc cutters on a full-face tunnelling machine. *Symposium 'Tunnelling '82', Inst.Min.Met., London*: 127-133.

Larson, D.A., R.J. Morell & J.F. Mades 1987. An Investigation of Crack Propagation With a Wedge Indenter To Improve Rock Fragmentation Efficiency. Bureau of Mines Report of Investigations ; 9106 Minneapolis (MN) (USA) 1987.

Larson,D.A., R.J. Morell & D.E. Swanson 1987. Large Scale Testing of the Ripper Fragmentation System. Bureau of Mines Report of Investigations ; 9123 Minneapolis (MN) (USA) 1987.

Loes, L. 1971. Loose abrasive operations. *Abrasives Springer Verlag Wien*: 24-28, New York.

Lowe, P.T. & L.B. McQueen 1990. Ground Conditions and Construction Methods in the Malabar Ocean Outfall and Observations on Rock Cuttability. *VII Australian Tunnelling Conference "The Underground Domain", Preprints of Papers, Sydney*. The Institution of Civil Engineers, Australia.

MacGregor, F., R. Fell, G.R. Mostyn, G. Hocking & G. McNally 1994. The Estimation of Rock Rippability. *Quarterly J. of Eng. Geology*, vol. 27.

Malkin, S. 1989 *GRINDING TECHNOLOGY,Theory and Applications of Machining with abrasives*. Ellis Horwood Series in Mechanical Engineering, Chicester England.

Martin, J.M. 1986. Predicting the rippability of sandstone in SE Queensland. 13th ARRB/5th REAAA Conf: 119-132 UITGEVER

McFeat-Smith, I. & R.J. Fowell 1977. Correlation of rock properties and the cutting performance of tunnelling machines. *Proc. Conf. Rock Engineering, University Newcastle upon Tyne*: 587-602 The University of Newcastle upon Tyne, UK.

Mecklenburg, K.R. & R.I Benzing 1976. *Testing for Adhesive Wear, Selection and Use of Wear Tests for Metals ASTM STP 615*: 12-29 American Society of Testing Materials, Baltimore

Meer, J. van der 1979. Macroscopisch en microscopisch onderzoek van de gesteenten

Euville, Felser, Savonieres, WLII, Sirieul, St Leu, mergel hard, mergel zacht en gips. BAGT rapport no.297, Waterloopkundig Laboratorium, Delft.

Mishra, I. & I. Finnie 1979. A Classification of Three-Body Abrasive Wear and Design of A New Tester. *Wear* vol.60: 111-121.

Mishra, I. & I. Finnie 1981. Correlations between two-body and three-body abrasion and erosion of metals. *Wear* vol.68: 33-39.

Muro, T., 1978 Characteristics of Shape Variation of Rippertip by Wear and Abrasiveness of Rock. *Proc. of JSCE*, no. 274: 119-130.

Nishimatsu, Y., 1972. The mechanics of rock cutting. *Int. J. Rock Mech. Min. Sci. & Geomech. Abstr.*, vol. 9: 261-270.

Nishimatsu, Y., 1979. On the Effect of Tool Velocity in Rock Cutting *Int. Conference on Mining and Machinery, Brisbane.*

Osburn, H.J., 1969. Wear of Rock-Cutting Tools. *Powder Metallurgy* vol. 12, no. 24.

Paschen, D., 1980. Petrographische und geomechanische Charakteriserung von Ruhrkarbongesteinen zur Bestimmung ihres Versleissverhaltens. Clausthal, Dissertation, Technische Universität Clausthal.

Pells, P.J.N., 1985. Engineering properties of the Hawkesbury Sandstone. In P.J.N. Pells (ed.): *Engineering Geology of the Sydney Region*: 179-197, Balkema, Rotterdam.

Phillips, H.R. & F.F.Roxborough 1981. The influence of tool material on the wear rate of rock cutting picks. University of New South Wales, Australia.

Powers, M.C., 1953, A new roundness scale for sedimentary particles. *Journal of Sedimentary Petrology*, vol. 23, No. 2: 117-119.

Reinhardt H.W., 1985. *BETON als constructiemateriaal eigenschappen en duurzaamheid.* Delftse Universitaire Pers, Delft.

Roos, G.A.B., 1991. The Effects of Pressure and Temperature on Acoustic Waves through Porous Media. MSc. thesis Delft University of Technology, Faculty of Mining and Petroleum Engineering, Section Petrophysics, December 1991.

Rossen H.P.H., van 1987. Analyse van snij- en slijtageprocessen van beitels bij het snijden van vaste gesteenten. Rapportnummer 87.3.GV.2232, Technische Universiteit Delft, Faculteit der Werktuigbouwkunde, Techniek van het Grondverzet.

Roxborough, F.F, 1987. The role of some basic rock properties in assessing cuttability. *Proc. Seminar Tunnels-Wholly Engineered Structures, I.E. Aust/A.F.C.C.*, Sydney Australia.

Roxborough, F.F., 1973. Cutting Rock with Picks. *Mining Engineer*: 445-455.

Roxborough, F.F. & A. Rispin 1973 The Mechanical Cutting Characteristics of the Lower Chalk. *Tunnels and Tunnelling*: 45-67.

Roxborough, F.F. & G.C. Sen 1986. Breaking Coal and Rock. *Australian Coal Mining Practise AIMM*, Monograph Series Wo12.

Roxborough, F.F. & H.R. Philips 1981. The influence of tool material on the wear rate of rock cutting picks. University of New South Wales, Australia, 1981.

Schimazek, J. & H. Knatz 1970. Der Einfluss des Gesteinaufbaus auf die Schnittgeschwindigkeit und den Meisselversleiss von Streckenvortriebmaschinen. *Glückauf* Vol.106.6.: 274-278.

Schimazek, J. & H. Knatz 1976. Die Beurteilung der Bearbeitbarkeit von Gesteinen durch Schneid- und Rollenbohrwerkzeuge. *Erzmetall.* Vol. Bd. 29: 113-129.

Sellami, H. 1993 Prospective applications of pick cutting machines in agressive rocks: Feasability of activated cutting tools. *Mine Mechanization and Automation. Proc. of the 2nd International Symposium on Mine Mechanization and Automation, Luleå Sweden*: 287-295. Balkema, Rotterdam

Sman van der, R.M. 1989. Resultaten van de eerste series beitelslijtagemetingen m.b.v. de schaafbank. CSB Beitelslijtage, BAGT Bewerken van rots, 89.3.GV.2667., Delft.

Speight, H.E. & R.J. Fowell 1984. The influence of operational parameters of roadheader productivity and efficiency with particular reference to cutting pick wear. *IMechE.*

Steen, M.A.J.van der 1985. Oppervlakteruwheidsmeting II*. *Mikroniek*,25,5: 19-22.

Uetz,H., 1986. *Abrasion und Erosion* Carl Hauser Verlag München.

Verhoef, P.N.W., 1994. Wear testing categories for rock dredging projects. *Proc. of the 6th Int. Congress of the Int. Ass. of Eng. Geol., Amsterdam*: 3289-3295 Balkema, Rotterdam.

Verhoef, P.N.W., 1995. Wear of Rock Cutting Tools, Site Investigation for Rock Dredging., in preparation.

Verhoef, P.N.W., 1993. Abrasivity of Hawkesbury Sandstone (Sydney, Australia) in relation to rock dredging. *Quarterly Journal of Engineering Geology*, vol. 26: 5-17.

Verhoef, P.N.W., H.J. van den Bold & Th.W.M. Vermeer, 1990. Influence of microscopic structure on abrasivity of rock as determined by the pin-on-disc test. *Proc. 6th Int.Congress of the Int. Ass. of Eng. Geol., Amsterdam*: 495-405 Balkema, Rotterdam.

Weaver, J.M., 1975. Geological Factors Significant in the Assessment of Rippability. *The civil Engineer in South Africa*, 17: 313-316.

West, G., 1982. A review of rock abrasiveness testing for tunnelling. *Proceedings of the International Symposium on Weak Rock, Tokyo*: 585-594 Balkema, Rotterdam.

Whittaker, B.N. & A.B. Szwilski, 1973 Rock cutting by impact action. *Int. J. Rock Mech. Min. Sci. & Geomech. Abstr.* Vol. 10: 659-671.

Zum Gahr, K.H., 1987. *Microstructure and Wear of Materials*. Tribology Series 10, Elsevier, Rotterdam.

Comparison of porosity values determined buy different methods

Rock property parameters, affecting the rate and type of wear, depend often upon the method of determination. Thus the influence of the test method may have a significant influence on the final wear prediction, which is based on such rock property parameters. Most rock property parameters in this research are determined by one rock property parameter assessment method only. The composition of the artificial and natural rock types however are differently determined. The composition of the artificial rocks has been deducted from the contents of the mortar paste when the mortar was casted, while the composition of the natural rocks has been determined by thin section analysis.

Porosity for example can be determined in several ways. In case of mortar the exact volume percentages of the different ingredients, mixed during casting, is known. The density of the mortar material can be determined from these data or the specific gravity can be determined with the density bottle method (BS 1377:1975, Test 6(B)). The porosity can then be calculated from the mortar density or the specific gravity value and the dry density (ASTM C97-83). Porosity can also be determined from thin section analysis (point counting method). Table A shows porosity values of some mortar compositions, obtained by the different assessment methods.

Table A

Mortar Composition	Porosity %		
	calculated from mortar material density and dry density	calculated from the specific gravity and dry density	determined by thin section analysis
mc 1	10	18	21
mc 2	18	25	17
mc 3	10	18	13
mc 4	11	24	15
mc 5	9	20	16
mc 6	19	27	5
mc 7	7	14	6
mc 8	24	37	17
mc 9	9	20	8

APPENDIX II

Petrographic descriptions (BS 5930 : 1981)

PETROGRAPHIC DESCRIPTION Schnarrtanne granite

name:M.Giezen **date**.Nov.1993 **location**:Schnarrtanne, Germany **geological formation**:Coarse grained tourmaline granite of Eibenstock.	**macroscopic description**: Pinkish white, coarse grained, wholly discoloured, mineralogically altered, micro-fractured, tourmaline GRANITE, moderately strong.

mineralogy

thin section numbers:	point-count:1500	mineral:	vol.%:	average grain size:
0,00,000		quartz	30.7	1-2 mm
		alkali-feldspars	42.7	
		plagioclase	18.9	
		biotite	4.7	
		muscovite	3.0	

general remarks: The granite has a crystalline texture with minerals of an average size ranging from 2 to 4 mm.
The minerals show typical signs of weathering. A lot of feldspars show (micro) cracks, in which micas have been crystallized replacing the feldspars (sericite). The alkali-feldspars have severely been kaolinized. The plagioclase minerals are partly transformed to alkali-feldspars (anti-pertite). Even (weathering-resistant) quartz minerals show micro-cracks.

petrographic classification: granite

Enlargement approx: 8x ; crossed nicols: no

PETROGRAPHIC DESCRIPTION Oberndorfer sandstone				
name:M.Giezen date:Nov.1993 location:Oberndorf, Germany. geological formation:Lower Buntsandstein (Triassic).	**macroscopic description**:Yellowish grey, medium grained, partially discoloured, calcareous SANDSTONE, moderately strong.			
mineralogy				
thin section numbers:	point-count:800	mineral:	vol.%:	average grain size:
3C,1A,2B		quartz	43.5	0.4 mm
		calcite	36.0	
		alkali-feldspars	15.5	
		rock-fragments	1.2	
		iron-hydroxide	2.9	
		micas	0.9	

general remarks: The sandstone exists of a fine-grained calcite ground matrix with larger grains and minerals in it. Part of the fine-grained calcites is transferred into larger minerals by a re-crystallization process. The average grain size is about 0.25 mm. The shape of the quartz grains according to Powers' scale (1953) is "sub-rounded" with a high sphericity.

petrographic classification: calcareous sandstone

Enlargement approx: 8x ; crossed nicols: yes

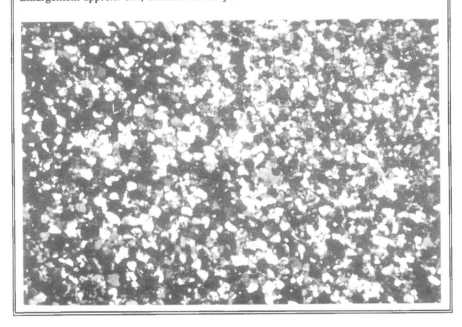

PETROGRAPHIC DESCRIPTION Bentheimer sandstone				
name·H.I.R.Deketh date:March 1994 location:Germany. geological formation:		macroscopic description:Yellowish grey, fine grained, wholly discoloured feldspars, SANDSTONE, strong.		
mineralogy				
thin section numbers:	point-count:500	mineral:	vol.%:	average grain size:
B		quartz	84.4	0.164
		feldspars	3.0	
		clay minerals	6.0	
		opaque minerals	0.8	
		lithic material	5.0	

general remarks: The sandstone exists of monocrystalline quartz grains with quartz overgrowth, which often binds the grains together. The feldspars were often heavily weathered. The shape of the quartz grains according to Powers' scale (1953) is "sub-angular". Grain sizes ranged from 0.054 to 0.36 mm.

petrographic classification: quartz arenite

enlargement approximately 6x

plane polarized light   crossed nicols

PETROGRAPHIC DESCRIPTION Felser sandstone				
name:H.J.R.Deketh **date**:March 1994 **location**: Germany **geological formation**:		**macroscopic description**:Reddish, medium grained, wholly discoloured feldspars, SANDSTONE, moderately strong.		
mineralogy				
thin section numbers:	point-count:500	mineral:	vol.%:	average grain size:
A		quartz	47	0.272
		pot. feldspars	12.4	
		clay minerals	15.6	
		opaque minerals	1.6	
		lithic material	8.2	
		biotite	12.0	
		muscovite	3.0	

general remarks: The feldspars were heavily weathered and often altered to clay minerals. The shape of the quartz grains according to Powers' scale (1953) is "angular". The grain sizes range from 0.09 to 0.54 mm.

petrographic classification: feldspatic greywacke

enlargement approximately 6x

plane polarized light   crossed nicols

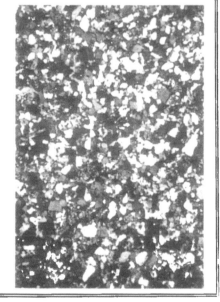

PETROGRAPHIC DESCRIPTION Elb sandstone D

name: H.J.R. Deketh date:March 1994 location: Germany	macroscopic description:Pinkish, medium grained, fresh, SANDSTONE, strong.
geological formation:	

mineralogy

thin section numbers:	point-count:500	mineral:	vol.%:	average grain size:
D		quartz	91.2	0.35
		clay,mica	8.8	

general remarks: The shape of the quartz grains according to Powers' scale (1953) is "sub-angular". The grain sizes range from 0.3 to 2 mm mm.

petrographic classification: quartz arenite

enlargement approximately 6x

plane polarized light   crossed nicols

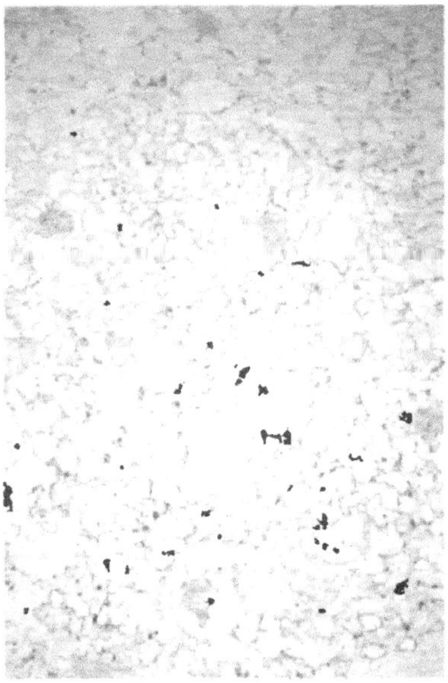

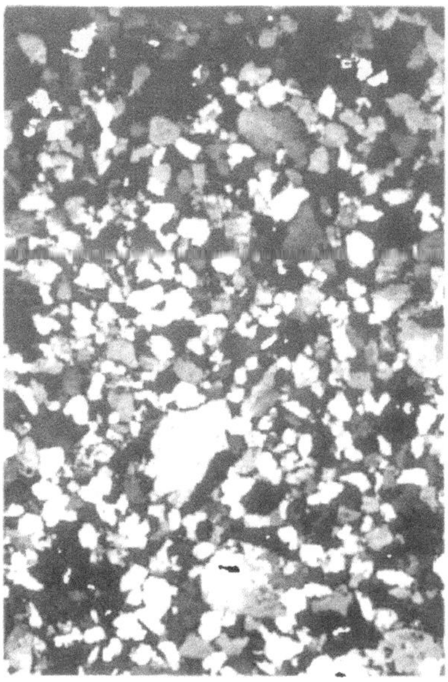

PETROGRAPHIC DESCRIPTION Elb sandstone C				
name:H.J.R.Deketh **date**:March 1994 **location**: Germany		**macroscopic description**:White, fine grained, fresh, SANDSTONE, moderately strong.		
geological formation:				
mineralogy				
thin section numbers:	point-count:500	mineral:	vol.%:	average grain size:
C		quartz	76.6	0.176
		feldspar	7.6	
		clay minerals	7.8	
		opaque minerals	0.6	
		lithic material	5.4	
		zircon	1.4	
		tourmaline	0.6	

general remarks: The shape of the quartz grains according to Powers' scale (1953) is "sub-angular". The grain sizes range from 0.054 to 0.54 mm.

petrographic classification: quartz arenite

enlargement approximately 6x

plane polarized light   crossed nicols

PETROGRAPHIC DESCRIPTION sandstone T

name:H.J.R.Deketh date:March 1994 location: Czecho Slovakia	macroscopic description:Pinkish, medium grained, SANDSTONE, moderately strong.
geological formation:	

mineralogy

thin section numbers:	point-count:500	mineral:	vol.%:	average grain size:
T		quartz	64.5	0.176
		feldspar	10.1	
		clay minerals	10.3	
		opaque minerals	2.9	
		lithic material	12.2	

general remarks: The shape of the quartz grains according to Powers' scale (1953) is "angular to sub-angular". The grain sizes range from 0.140 to 1.40 mm.

petrographic classification: sub-arkose

enlargement approximately 6x

plane polarized light   crossed nicols

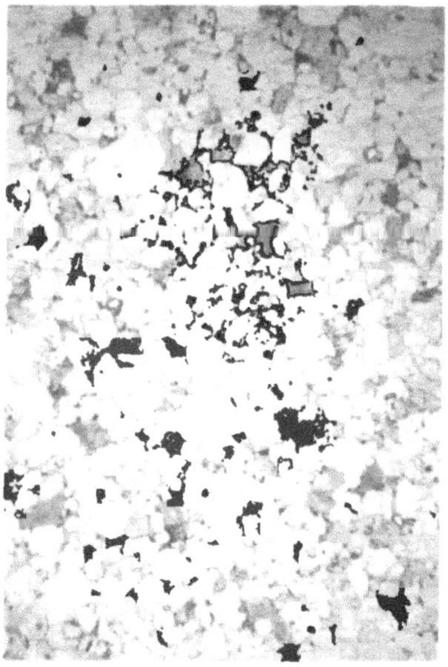

PETROGRAPHIC DESCRIPTION Euville

name:H.J.R.Deketh date:March 1994 location:	**macroscopic description**: Mottled white, coarse grained, CALCARENITE, moderately strong.
geological formation:	

mineralogy

thin section numbers:	point-count:500	mineral:	vol.%:	average grain size:
EUNNII		calcite grains	95.0	1.60
		calcite cement	5.0	
			.	

general remarks: The grain sizes range from 1.00 to 40.0 mm. Calcite grains consists mainly of shell debris of various sizes. The cement is mainly very fine calcite mud, sometimes recrystallized.

petrographic classification: calcarenite

enlargement approximately 6 x

plane polarized light   crossed nicols

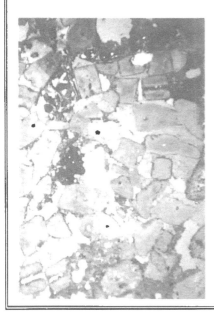
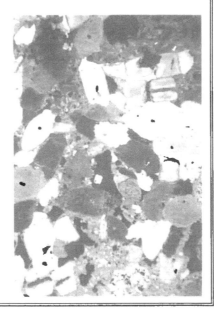

PETROGRAPHIC DESCRIPTION Sirieul				
name:H.J.R.Deketh **date**:March 1994 **location**:		**macroscopic description**: White, fine grained, LIMESTONE, moderately weak.		
geological formation:				
mineralogy				
thin section numbers:	point- count:500	mineral:	vol. %:	average grain size:
EUNNII		calcite peloids	71.4	
		calcite cement	16.6	
		shell fragments	7.0	
		quartz	5.0	0.161

general remarks: The shape of the quartz grains according to Powers' scale (1953) is "sub-angular". The grain sizes range from 0.36 to 0.45 mm. The peloids are micritisized.

petrographic classification: Pelmicrite

enlargement approximately 6x

plane polarized light   crossed nicols

APPENDIX III

Momentary cutting forces in tests on Elb-sandstone D.

The cutting forces in each test run vary with the displacement or time. In this work the average forces were used to describe the cutting process. However the average forces do not indicate the variation of the forces in a test. In figure A, the momentary cutting forces (Fz (cutting force) and Fy (normal force)) in a representative time period of a test with a low feed (0.07 mm per revolution) on Elb-sandstone D is shown. Momentary cutting forces in a test at a higher feed (0.57 mm per revolution) on the same rock type and during the same time period as in Figure A is shown in Figure B. The test at the small feed took place in mode I and the test at the larger feed in mode III. The variation of the cutting forces around the average is higher for the test at the larger feed (Figure B). The standard deviation of the forces in Figure A is about 28% of the average cutting forces and in Figure B about 36% of the average cutting forces. Thus, the vibration in the cutting apparatus is likely to increase with an increase of the feed. The saw-tooth pattern of the momentary cutting forces as observed in rock cutting experiments by Bisschop (1990) and Davids and Adrichem (1990) could not be seen in the scraping test experiments. Either even the largest feed in these tests is still too small to produce rock chips of considerable size or the sample frequency of the cutting forces (5 readings per second) is not high enough to reveal the saw-tooth pattern.

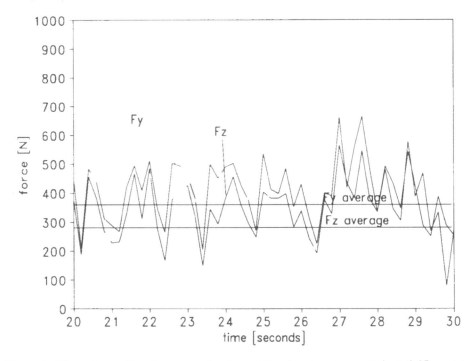

Figure A. Momentary cutting forces as a function of time in a test at a small feed, 0.07 mm/rev. (MODE I) at a cutting velocity of 0.4 m/s

143

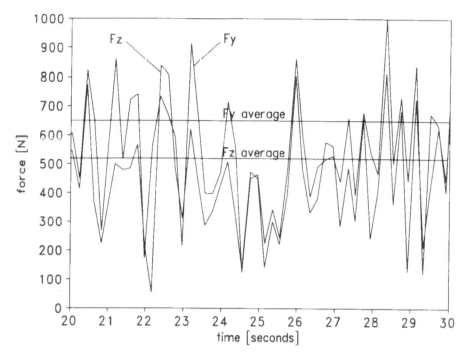

Figure B. Momentary cutting forces as a function of time in a test at a large feed, 0.57 mm/rev. (MODE III) at a cutting velocity of 0.4 m/s

T - #0272 - 101024 - C0 - 254/178/9 [11] - CB - 9789054106203 - Gloss Lamination